Applied Data Communications and Networks

Applied Data Communications and Networks

W. Buchanan
Senior Lecturer
Department of EECE
Napier University
Edinburgh
UK

DON PELTON

CHAPMAN & HALL

London · Weinheim · New York · Tokyo · Melbourne · Madras

Published by Chapman & Hall, 2–6 Boundary Row, London SE1 8HN, UK

Chapman & Hall, 2–6 Boundary Row, London SE1 8HN, UK

Chapman & Hall GmbH, Pappelallee 3, 69469 Weinheim, Germany

Chapman & Hall USA, 115 Fifth Avenue, New York, NY 10003, USA

Chapman & Hall Japan, ITP-Japan, Kyowa Building, 3F, 2-2-1 Hirakawacho, Chiyoda-ku, Tokyo 102, Japan

Chapman & Hall Australia, 102 Dodds Street, South Melbourne, Victoria 3205, Australia

Chapman & Hall India, R. Seshadri, 32 Second Main Road, CIT East, Madras 600 035, India

First edition 1996

© 1996 Chapman & Hall

Printed in Great Britain by The Alden Press, Osney Mead, Oxford

ISBN 0 412 75430 4

A catalogue record for this book is available from the British Library

This book is dedicated to John and Margaret whose kindness and help over the years has been worth more than money can buy.

Contents

Preface

The usage of data communications and computer networks are ever increasing. It is one of the few technological areas which brings benefits to most of the countries and the peoples of the world. Without it many industries could not exist. It is the objective of this book to discuss data communications in a readable form that students and professionals all over the world can understand. As much as possible the text uses diagrams to illustrate key points.

Most currently available data communications books take their viewpoint from either a computer scientists top-down approach or from an electronic engineers bottom-up approach. This book takes a practical approach and supports it with a theoretical background to create a textbook which can be used by electronic engineers, computer engineers, computer scientists and industry professionals.

It discusses most of the current and future key data communications technologies, including:

- Data Communications Standards and Models;
- Local Area Networks (Ethernet, Token Ring and FDDI);
- Transmission Control Protocol/Internet Protocol (TCP/IP);
- High-level Data Link Control (HDLC);
- X.25 Packet-switching;
- Asynchronous Communications (RS-232) and Modems;
- Pulse Coded Modulation (PCM);
- Integrated Digital Services Network (ISDN);
- Asynchronous Transfer Mode (ATM);
- Error Control;
- X-Windows.

The chapters are ordered in a possible structure for the presentation of the material and have not been sectioned into data communications areas.

There are six main areas covered and these can be linked with the chapters to give the following:

- Introduction to communications systems (chapters 1 and 2);
- LANs and internetworking (chapters 3, 4 and 5);
- WANs (chapters 7, 8, 9, 10 and 11);
- Current and future provision (chapters 12 and 17);
- Physical layer - signalling and media (chapters 15, 16 and 17);
- Related/ pervasive issues (chapters 6 and 13).

One of the main objects of the book is to use practical examples to illustrate key areas. These include:

- the analysis of a real-life university Ethernet and Token Ring computer network, showing how interconnected networks communicate;
- the analysis of state-of-the-art technologies such as FDDI and ATM with reference to a practical Metropolitan Area Network;
- practical data communications programs, in C, for serial communications;
- practical examples of communications over the Internet;
- the necessary information on how to set-up computers for TCP/IP communications.

The text also contains some multiple choice exercises which can be used by a lecturer to revise the material presented or can be used at the beginning of a tutorial session.

Requests for technical support can be sent to the author using the email address:

```
w.buchanan@central.napier.ac.uk
```

Book updates, together with the software in the book, and other related information can be found on the WWW page:

```
http://www.eece.napier.ac.uk/~bill_b/dbook.html
```

Finally, I would like to thank my commissioning editor Dave Hatter for his support of this project and for his dynamic energy. I would also like to thank Ian Marshall, in the School of Information Systems at University of East Anglia, for his expert knowledge and comments on the original proposal. His feedback was most helpful during the writing of this book.

1

Communications systems

1.1 INTRODUCTION

The usage of electronic communications increases by the day. In the past, most electronic communication systems transmitted analogue signals. On an analogue telephone system the voltage level from the phone varies with the voice signal. Unwanted signals from external sources easily corrupt these signals. In a digital communication system a series of digital codes represents the analogue signal. These are then transmitted as 1's and 0's. Digital information is less likely to be affected by noise and has thus become the most predominent form of communications.

Digital communication also offers a greater number of services, greater traffic and allows for high speed communications between digital equipment. The usage of digital communications includes cable television, computer networks, facsimile, mobile digital radio, digital FM radio and so on.

1.2 COMMUNICATIONS MODEL

Figure 1.1 shows a communications model in its simplest form. An information source transmits signals to a destination through a transmission media. Unfortunately, these signals may not be in a transmittable form for the transmission media. For example, a person in Japan could not communicate with another in France by simply shouting a message as the sound waves produced would soon die away. A better solution would be to transmit the signal using low frequency radio waves. These waves propagate through air better than sound waves.

In this case the radio wave is defined as the carrier wave as it transports the information signal. The modulated wave is the result of the carrier and the information signal. After transmission through the medium the modulated wave is demodulated by the receiver. Figure 1.2 shows an improved communications model with a modulator at the source and a demodulator at the receiver.

A modulated wave also allows many sources to transmit over the same media, at the same time. The receiver selects the source information by 'tuning' into the required carrier wave.

Unfortunately, noise is also added to the transmitted signal. Noise is any unwanted signal and has many causes, such as static pick-up, poor electrical connections, electronic noise in components, cross-talk, and so on. It makes the reception of a signal more difficult and can also produce unwanted distortion on the unmodulated signal.

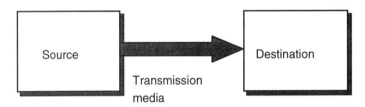

Figure 1.1 Simple communications model

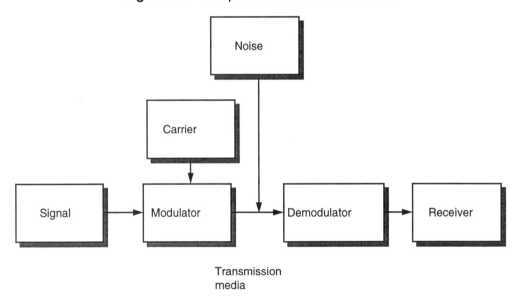

Figure 1.2 Communications model

1.3 ELECTRONIC SIGNALS

Any electrical signal can be analyzed either in the time-domain or in the frequency-domain. A time-varying signal contains a range of frequencies.

If the signal is repetitive (that is, it repeats after a given time) then the frequencies contained in it will also be discrete. This will be discussed in more detail in Chapter 15.

The standard form of a single frequency signal is:

$$V(t) = V \sin(2\pi ft + \theta)$$

where *v(t)* is the time varying voltage (V), *V* is the peak voltage (V), *f* the signal frequency (Hz) and θ its phase (°)

A signal in the time-domain is a time varying voltage. In the frequency domain it is voltage amplitude against frequency. Figure 1.3 shows how a single frequency is represented in the time-domain and the frequency domain. It shows that for a signal with a period T that the frequency of the signal is 1/T Hz. The signal frequency is represented in the frequency domain as a single vertical arrow at that frequency. The amplitude of this arrow represents the amplitude of the signal.

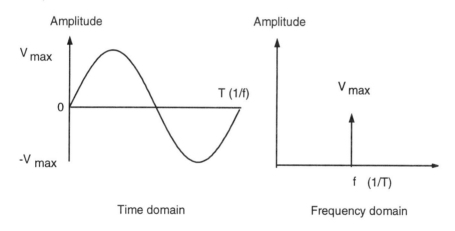

Figure 1.3 Representation of signal in frequency and time domains

1.4 BANDWIDTH

Bandwidth is the range of frequencies contained in a signal. As an approximation it is the difference between the highest and the lowest signal frequency, as illustrated in Figure 1.4. For example, if a signal has an upper frequency of 100 MHz and a lower of 75 MHz then the signal bandwidth is 25 MHz. Normally, the larger the bandwidth the greater the information sent. Unfortunately, normally, the larger the bandwidth the more noise that is added to the signal. The bandwidth of a signal is normally limited to reduce the amount of noise and to increase the number of signals transmitted. Table 1.1 shows typical bandwidths for different signals.

The two most significant limitations on a communication system performance are noise and bandwidth.

Table 1.1   Typical signal bandwidths

Application	*Bandwidth*
Telephone speech	4 kHz
Hi-fi audio	20 kHz
FM radio	200 kHz
TV signals	6 MHz
Satellite communications	500 MHz

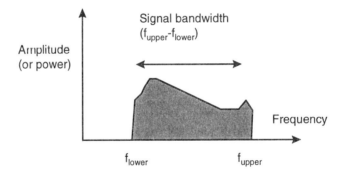

Figure 1.4   Signal bandwidth

1.5 SIGNAL FREQUENCY CONTENT

The greater the rate of change of an electronic signal the higher the frequencies that will be contained in the frequency response. Figure 1.5 shows two repetitive signals. The upper signal has a DC component and four frequencies, f_1 to f_4. The lower signal has a greater rate of change than the upper signal and it thus contains a higher frequency content, from f_1 to f_6.

Digital pulses have a very high rate of change around their edges. Thus, digital signals normally have a larger bandwidth than analogue signals.

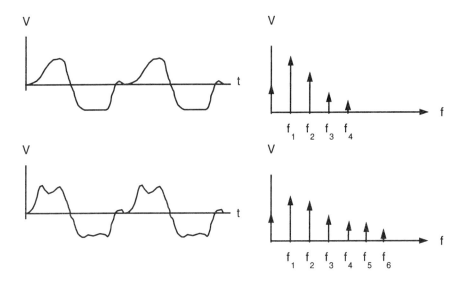

Figure 1.5 Frequency content of two repetitive signal

1.6 MODULATION

Modulation allows the transmission of a signal through a transmission medium by means of adding a carrier wave. It also adds extra information that allows the receiver to pick-up the signal.

There are three main methods used to modulate: amplitude, frequency and phase modulation. With amplitude modulation (AM) the information signal varies the amplitude of a carrier wave. In frequency modulation (FM) it varies the frequency of the wave and with phase modulation (PM) it varies the phase.

1.6.1 Amplitude modulation (AM)

AM is the simplest form of modulation where the information signal modulates a higher frequency carrier. The modulation index, m, is the ratio of the signal amplitude to the carrier amplitude. It is always less than or equal to 1 and is given by:

$$m = \frac{V_{signal}}{V_{carrier}}$$

Figures 1.6 to 1.8 show 3 differing modulation indices. In Figure 1.6 the information signal has a relatively small amplitude compared with the carrier signal. The modulation index, in this case, is relatively small.

In Figure 1.7 the signal amplitude is approximately half of the carrier amplitude. In Figure 1.8 the signal amplitude is almost equal to the carriers amplitude. The modulation index in this case will be near unity.

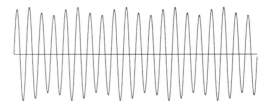

Figure 1.6   AM waveform

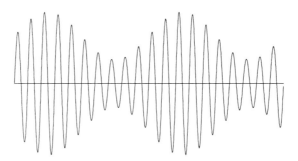

Figure 1.7   AM waveform

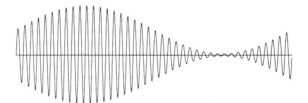

Figure 1.8   AM waveform

AM is susceptible to noise and fading as it is dependent on the amplitude of the modulated wave.

1.6.2 Frequency modulation (FM)

Frequency modulation involves the modulation of the frequency of a carrier. FM is preferable to AM as it is less affected by noise because the information is contained in the change of frequency and not the amplitude. Thus, the only noise that affects the signal is limited to a small band of frequencies contained in the carrier. The information in an AM waveform is contained in its amplitude which can be easily affected by noise.

Figure 1.9 shows a modulator/ demodulator FM system. A typical device used in FM is a Phased-Locked Loop (PLL) which converts the received frequency-modulated signal into a signal voltage. It locks onto frequencies within a certain range (named the capture range) and follows the modulated signal within a given frequency band (named the lock range).

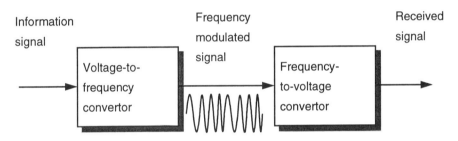

Figure 1.9 Frequency modulation

FM radio broadcasting requires a much larger signal bandwidth than AM. Typically, FM radio transmits from 88 to 108 MHz. The peak frequency deviation in each signal of 75 kHz (that is, the lower frequency is the carrier less 75 kHz and the upper is the carrier plus 75 kHz). Thus, the signal bandwidth for FM is at least 150 kHz.

1.6.3 Phase modulation (PM)

Phase modulation involves the modulation the phase of the carrier. PM is less affected by noise than AM because the information is contained in the change of phase and, like FM, not in its amplitude.

1.6.4 Digital modulation

To send digital signals over a limited bandwidth channel, such as a speech channel, the digital pulses may have to be changed into frequencies that are within the bandwidth of the channel. This can be achieved either using amplitude, frequency or phase modulation.

Frequency-shift keying (FSK) uses two frequencies to transmit a 1 and a 0, that is, an upper frequency and a lower frequency. In phase-shift keying (PSK) a 1 is transmitted with no change in phase, else an amplitude inversion (180° phase shift) corresponds to a 0. With amplitude-shift keying (ASK) the amplitude of the carrier is varied in one of two states. Figure 1.10 shows an example of the transmission of the bit pattern 11010 for the three techniques.

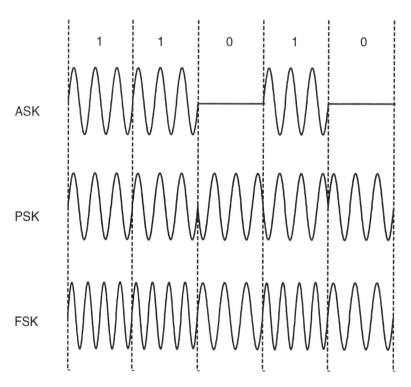

Figure 1.10   Frequency modulation

1.7 FREQUENCY CARRIER

The frequency carrier of a signal is important and is chosen for several reasons, such as the:

- signal bandwidth;
- signal frequency spectrum;
- transmission channel characteristics.

Figure 1.11 shows the frequency spectrum of electromagnetic (EM) waves. The microwave spectrum is sometimes split into millimetre wave and microwaves and the radio spectrum splits into 7 main bands from ELF (used for very long distance communications) to VHF (used for FM radio).

Normally, radio and lower frequency microwaves are specified as frequencies. Whereas, EM waves from high frequency (millimetre wavelength) microwaves upwards are specified as a wavelength.

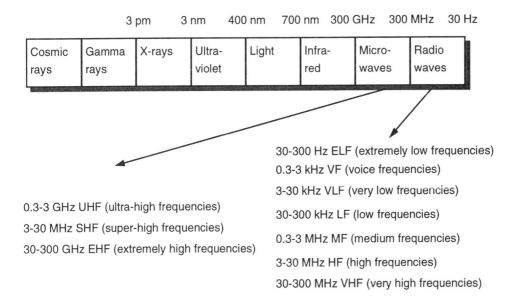

Figure 1.11   EM frequency spectrum

The wavelength of a signal is the ratio of its speed of propagation (u) to its frequency (f). It is thus given by:

$$\lambda = \frac{u}{f}$$

In free space an electromagnetic wave propagates at the speed of light ($300\,000\,000$ m.s^{-1} or $186\,000$ miles.s^{-1}). For example, if the carrier frequency of an FM radio station is 97.3 MHz then its transmitted wavelength is 3.08 m. If an AM radio station transmits at 909 kHz then the carrier wavelength is 330 m. Typically, the length of radio antennas is designed to be half the wavelength of the received wavelength. This is the reason why FM aerials are normally between 1 and 2 metres in length whereas in AM and LW aerials a long coil of wire is wrapped round a magnetic core. Note that a 50 Hz mains frequency propagates through space with a wavelength of 6 000 000 m.

If an EM wave propagates through a dense material then its speed slows. In terms of the dielectric constant, ε_r of a material (which is related to density) then the speed of propagation is:

$$u = \frac{c}{\sqrt{\varepsilon_r}}$$

Each classification of EM waves has its own characteristics. The main classifications of EM waves used for communication are:

- Radio waves: The lower the frequency of a radio wave the more able it is to bend round objects. Defence applications use low frequency communications as they can be transmitted over large distances, and over and round solid objects. The trade-off is that the lower the frequency the less the information that can be carried. LW (MF) and AM (HF) signals can propagate large distances, but FM (VHF) requires repeaters because they cannot bend round and over solid objects such as trees and hills. Long wave radio (LW) transmitters operate from approximately 100 to 300 kHz, medium wave (AM) from 0.5 to 2 MHz and VHF radio (FM) from 87 to 108 MHz.

- Microwaves: Microwaves have the advantage over optical waves (light, infra-red and ultra-violet) in that they can propagate well through water and thus can be transmitted thorough clouds, rain, and so on. If they are of a high enough frequency they can propagate through the ionosphere and out into outer space. This property is used in satellite communications where the transmitter bounces microwave energy off a satellite, which is then picked up at a receiving station. Radar and mobile radio applications use these properties. Their main disadvantage is that they will not bend round large objects as their wavelength is too small. Included in this classification is UHF (used to transmit TV signals), SHF (satellite communications) and EHF waves.

- Infra-red: Infra-red is used in optical communications. When used as a carrier frequency the transmitted signal can have a very large bandwidth because the carrier frequency is high. It is extensively used for line-of-site communications especially in remote control applications. Infra-red radiation is basically the propagation of heat. Heat received from the sun propagates as infra-red radiation.

- Light: Light is the only part of the spectrum that humans can 'see'. It is a very small part of the spectrum and ranges from 300 to 900 nm. Colours contained are Red, Orange, Yellow, Green, Blue, Indigo and Violet (ROY.G.BIV or **R**ichard **O**f **Y**ork **G**ave **B**attle **I**n **V**ain).

- Ultra-violet: As with infra-red it is used in optical communications. In high enough exposures it can cause skin cancer. The ozone layer blocks much of the ultra-violet radiation from the sun.

1.8 MULTIPLEXING

Multiplexing is a method of sending information from many sources over a single transmission media. For example, satellite communications and optical fibres allow many information channels to be transmitted simultaneously. There are two main methods of achieving this, either by separation in time with time-division multiplexing (TDM) or separation in frequency with frequency-division multiplexing (FDM).

1.8.1 Frequency-Division Multiplexing (FDM)

With FDM each channel uses a different frequency band. An example of this are FM radio and satellite communications. With FM radio many channels share the same transmission media but are separated into different carrier frequencies. Satellite communication normally involves an earth station transmitting on one frequency (the up-link frequency) and the satellite relays this signal at a lower frequency (the down-link frequency).

Figure 1.12 shows an FDM radio system where each radio station is assigned a range of frequencies for their transmission. The receiver then tunes into the required carrier frequency.

1.8.2 Time-Division Multiplexing (TDM)

With TDM different sources have a time slot in which their information is transmitted. The most common type of modulation in TDM systems is pulsed code modulation (PCM). With PCM, analogue signals are sampled and converted into digital codes. These are then transmitted as binary digits.

In a PCM-TDM system, several voice-band channels are sampled and converted into PCM codes. Each channel gets a time slot and each time slot is built up into a frame. The complete frame has extra data added to it to allow synchronization. Figure 1.13 shows a PCM-TDM system with three sources.

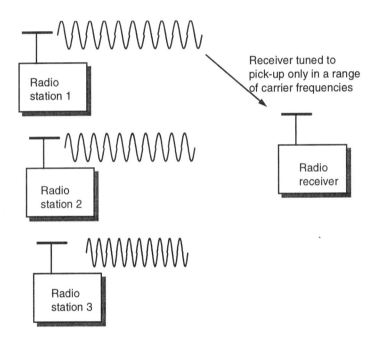

Figure 1.12   FDM radio system

1.9 NOISE AND SIGNAL DISTORTION

Noise is any unwanted signal added to information transmission. The main sources of noise on a communication system are:

- thermal noise;
- cross-talk;

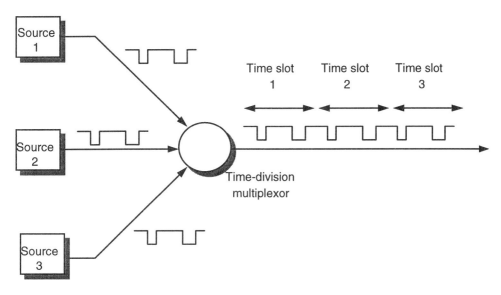

Figure 1.13   TDM system

* impulse noise.

Thermal noise arises from the random movement of electrons in a conductor and is independent of frequency. The noise power can be predicted from the formula:

$$N = k\,T\,B$$

where N is the noise power in Watts, k is Boltzman's constant (1.38 × 10^{-23} J/K) and B the bandwidth of channel (Hz). Thermal noise is predictable and is spread across the bandwidth of the system. It is unavoidable but can be reduced by reducing the temperature of the components causing the thermal noise. Many receivers which detect very small signals require to be cooled to a very low temperature in order to reduce thermal noise. A typical example is in astronomy where the receiving sensor is reduced to almost absolute zero. Thermal noise is a fundamental limiting factor in any communications system performance.

Impulse noise is any unpredictable electromagnetic disturbance, such as from lightning or from energy radiated from an electric motor. It is normally characterized by a relatively high energy, short duration pulse. It is of little importance to an analogue transmission system as it can usually be filtered out at the receiver. However, impulse noise in a digital system can cause the corruption of a significant number of bits.

Electrical signals propagate with an electric and a magnetic field. If two conductors are laid beside each other then the magnetic field from one can couple into the other. This is known as cross-talk, where one signal interferes with another. Analogue systems tend to be affected more by cross-talk than digital ones.

A signal can be distorted in many ways. The electrical characteristics of the transmitter and receiver and also the characteristics of the transmission media. An electrical cable contains inductance, capacitance and resistance. The inductance and capacitance have the effect of distorting the shape of the signal whereas resistance causes the amplitude of the signal to reduce (and also to lose power).

1.10 CAPACITY

The information-carrying capacity of a communications system is directly proportional to the bandwidth of the signals it carries. The wider the bandwidth the greater the information-carrying capacity. Typically, the maximum bandwidth of a system is 10% of the carrier frequency. Table 1.2 shows a few examples.

Table 1.2   Capacity

	Carrier	*Bandwidth (10%)*
VHF radio	100 MHz	10 MHz
Microwave	6 GHz	600 MHz
Light	10^{14} Hz	100 000 GHz

An important parameter for determining the capacity of a channel is the *signal-to-noise ratio* (SNR). This is usually given in decibels as the following:

$$\frac{S}{N}(dB) = 10 \log_{10} \frac{\text{Signal Power}}{\text{Noise Power}}$$

In a digital system, Nyquist predicted that the maximum capacity, in bits/sec, of a channel subject to noise is given by the equation:

$$\text{Capacity} = B.\log_2 \left[1 + \frac{S}{N}\right] \quad \text{bits / sec}$$

where B is the bandwidth of the system and S/N is the signal-to-noise ratio. For example if the signal-to-noise ratio is 10 000 and the bandwidth is 100 kHz, then the maximum capacity is:

$$\text{Capacity} = 10^5 . \log_2 \left[1 + 10^4 \right] \quad \text{bits / sec}$$

$$\approx 10^5 . \frac{\log_{10} \left[10^4 \right]}{\log_{10} [2]} \quad \text{bits / sec}$$

$$= 13.3 \times 10^5 \qquad \text{bits / sec}$$

1.11 TRANSMISSION MEDIA

Signals are transmitted over a suitable media which can affect the signal. Typical transmission media include:

• the earth's atmosphere (radio waves);
• optical fibres (light/infra-red);
• twisted-pair cable;
• coaxial cable;
• wave-guides (microwave transmission);
• micro-strip (microwave transmission and high-speed digital).

When a signal is transmitted over a media it will be affected by:

• attenuation;
• distortion;
• noise.

Attenuation is the loss of signal power and is normally frequency dependent. A low-pass channel is one which attenuates, or reduces, the high frequency components of the signal more than the low frequency parts. A band-pass channel attenuates both high and low frequencies more than a band in the middle.

The bandwidth of a system is usually defined as the range of frequencies passed which are not attenuated by more than half their original power level. The end points in Figure 1.14, are marked at 3 dB (the −3 dB point) above the minimum signal attenuation.

Bandwidth is one of the most fundamental factors as it limits the amount of information which can be carried in a channel at a given time. It can be shown that the maximum possible bit rate in a digital system on a noiseless, band-limited channel is twice the channel bandwidth, or:

$$\text{Maximum bit rate} = 2 \times \text{Bandwidth of channel}$$

If a signal is transmitted over a channel which only passes a narrow range of frequencies than is contained in the signal then the signal will be distorted.

If the frequency characteristics of the channel are known then the receiver can be given appropriate compensatory characteristics. For example, a receiving amplifier could boost higher frequency signals more than the lower frequencies. This is commonly done with telephone lines, where it is known as equalization.

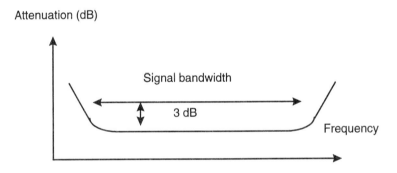

Figure 1.14   Bandwidth of a channel

1.12 EXERCISE

1.1   The frequency response of a signal is given in Figure 1.15, what is its bandwidth:

    A   20 kHz
    B   80 kHz
    C   100 kHz
    D   120 kHz

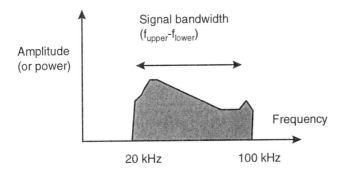

Figure 1.15   Signal frequency content

1.2   Which of the following best describes a modulator:

   A   it allows the information to be transmitted faster
   B   it allows the information to be transmitted slower
   C   it gives a reduction in the received noise
   D   it allows several channels to communicate over the same transmission media

1.3   What are the two most significant limitations on a communication system:

   A   transmitter and receiver design
   B   bandwidth and modulator design
   C   bandwidth and noise
   D   noise and the type of information transmitted

1.4   FM radio transmits within which range of frequencies:

   A   100–300 kHz
   B   0.5–1.6 MHz
   C   1–10 MHz
   D   88–108 MHz

1.5   AM radio transmits within which range of frequencies:

A   100–300 kHz
B   0.5–1.6 MHz
C   1–10 MHz
D   88–108 MHz

1.6   LW radio transmits within which range of frequencies:

A   100–300 kHz
B   0.5–1.6 MHz
C   1–10 MHz
D   88–108 MHz

1.7   Microwave EM waves have the advantage over optical waves in that they:

A   can carry more information
B   allow smaller transmitters to be designed
C   they propagate faster
D   can propagate through water better

1.8   Radio stations transmit signals with which of the following:

A   frequency-division multiplexing (FDM)
B   time-division multiplexing (TDM)
C   frequency-hopping multiplexing (FHM)
D   time-hopping multiplexing (THM)

1.9   Referring to Figure 1.16 which of the waveforms represent frequency-shift keying (FSK):

A
B
C
D

1.10 Referring to Figure 1.16 which of the waveforms represent amplitude-shift keying (ASK):

	✓
A	
B	
C	
D	

A
B
C
D

1.11 Referring to Figure 1.16 which of the waveforms represent phase-shift keying (PSK):

	✓
A	
B	
C	
D	

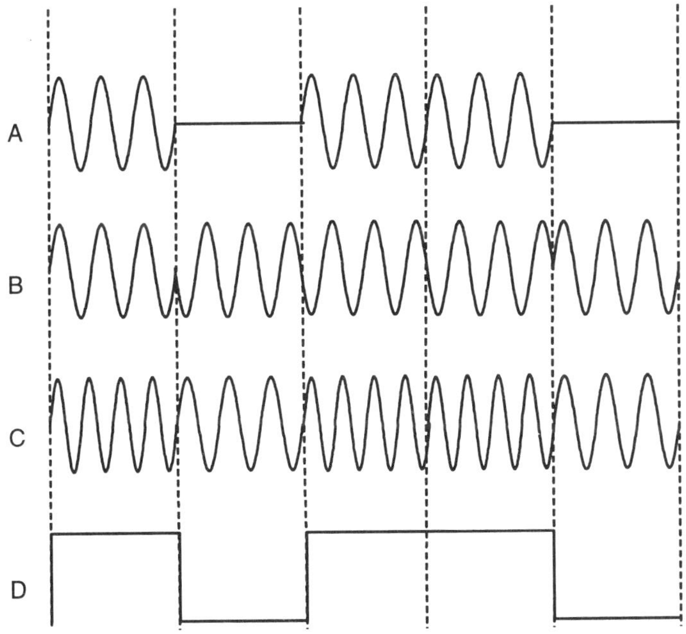

A
B
C
D

Figure 1.16 Digital modulation waveforms

2

Data communications models, networks and standards

2.1 COMMUNICATIONS NETWORKS

A local area network (LAN) is a collection of computers within a single office or building that connect to a common electronic connection – commonly known as a network backbone. A LAN can be connected to other networks either directly or through a WAN, as shown in Figure 2.1

A WAN normally connects networks over a large physical area, such as in different buildings, towns or even countries. Figure 2.1 shows four local area networks: LAN A, LAN B, LAN C and LAN D, some of which are connected by the WAN.

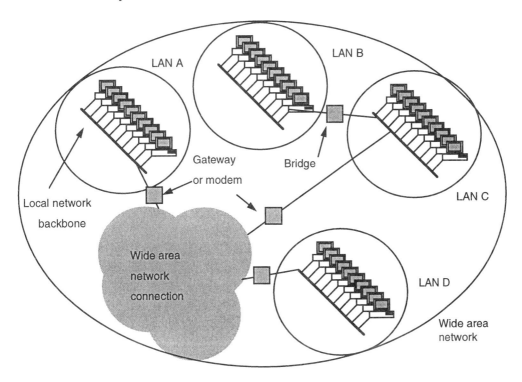

Figure 2.1   Interconnection of LANs to make a WAN

A modem connects a LAN to a WAN when the WAN connection is an analogue line. For a digital connection a gateway connects one type of LAN to another LAN, or WAN, and a bridge connects the same type of LAN to another.

The public switched telecommunications network (PSTN) provides long distance analogue lines. These public telephone lines can connect one network line to another using circuit switching. Unfortunately, they have a limited bandwidth and can normally only transmit frequencies from 400 to 3400 Hz. This results in a bandwidth of only 3 kHz. A modem is used to transmit digital signals over the PSTN and converts digital data into a transmittable form for the transmission line. Figure 2.2 shows the connection of computers to a PSTN.

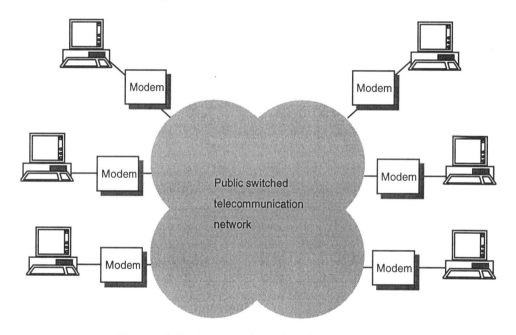

Figure 2.2   Connection of nodes to a PSTN

A public switched data network (PSDN) allows the direct connection of digital equipment to a digital network. This has the advantage of not requiring the conversion of digital data into an analogue form.

The integrated services digital network (ISDN) allows the transmission of many types of digital data into a truely global digital network. Data types include digitized video, digitized speech and computer data. Since the switching and transmission are digital, fast access times and relatively high bit-rates are possible. Typical base bit rates include 64 kbps (for digitized speech) and 16 kbps. All connections to the ISDN require network termination equipment (NTE). Figure 2.3 shows the connection of different types of digital equipment to an ISDN.

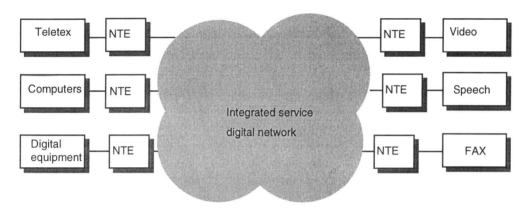

Figure 2.3   Connection of nodes to ISDN

2.2 LOCAL AREA NETWORKS

The following sections outline the advantages and disadvantages of local area networks.

2.2.1 Why network ?

Local area networks allow the orderly flow of information between connected nodes. Their main advantages are that:

- it is easier to set up new users and equipment;
- it allows the sharing of resources;
- it is easier to administer users;
- it is easier to administer software licences;
- it allows electronic mail to be sent between users;
- it allows simple electronic access to remote computers and sites;
- it allows the connection of different types of computers which can communicate with each other.

Figure 2.4 illustrates typical resources that can be set up with shared resources.

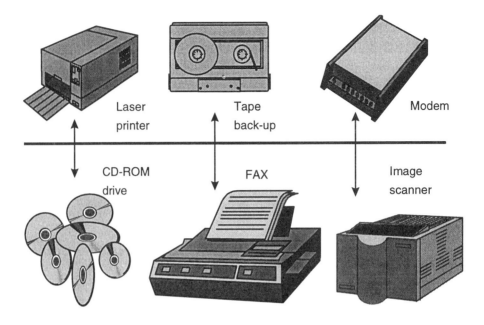

Figure 2.4   Local network with a range of facilities

2.2.2 Maintaining a network

A major advantage of LANs is their ability to share information over a network. Normally, it is easier to store application programs at a single location and make them available to users rather than to having copies individually installed on each computer (unless the application program requires special configurations or there are special licensing agreements). This saves on expensive disk space and increases the availability of common data and configurations. The disadvantage of this is that it increases the traffic on a network.

Most networks have a network manager, or network group, who manage the users and peripherals on a network. On a well-maintained network the network manager will:

- control the users on the network, that is, who can and cannot log in;
- control which of the users are allowed to use which facilities;
- control which of the users are allowed to run which application programs;
- control the usage of software packages by limiting users to licence

agreements;
- standardize the set up of application programs to a single source;
- back-up important files on a regular basis onto a mass back-up system;
- set up simple-to-use procedures to access programs, such as icons, menus, and so on;
- possibly control PC (Personal Computer) viruses by running automatic scanning programs;
- update application programs by modifying them at a single source.

2.2.3 Sharing resources

Computers not connected to a network may require extra peripherals such as printers, fax machines, modems, plotters, and so on. This may be resource inefficient as other users cannot get access to them unless they are physically disconnected and connected to their own computer. Normally, it is more efficient to share resources over a network.

Access to networked peripherals is also likely to be simpler as the system manager can standardize configurations. Peripherals that are relatively difficult to set up such as plotters, fax machines and modems are set up once and their configuration stored. The network manager can also bar certain users from using certain peripherals.

There is normally a trade-off between the usage of a peripheral and the number required. For example a single laser printer in a busy office may not be able to cope with the demand. A good network copes with this by segmentation, so that printers are assigned to different areas or users. The network may also allow for re-direction of printer data if a printer was to fail, or become busy.

2.2.4 Sharing disk resources (network file servers)

Many computer systems require access to a great deal of information and to run many application programs such as word processors, spreadsheets, compilers, presentation packages, computer-aided design (CAD) packages, and so on. Most local hard-disks could not store all the required data and application programs. A network allows users to access files and application programs on remote disks.

Multi-tasking operating systems such as Unix and VMS allow all hard-disks on a network to be electronically linked as a single file system.

Most PCs operating systems are not multi-tasking and are normally networked to a single file server computer. In this case the network server normally has a local drive that is available to all users on the network.

A network file server thus allows users access a central file system (for PCs) or a distributed file system (for Unix/ VMS). This is illustrated in Figure 2.5.

2.2.5 Electronic mail

Traditional methods of sending mail within an office environment are in-efficient as it normally requires an individual to request a secretary to type the letter. This must be proof-read and sent through the internal mail system which is relatively slow and can be open to security breaches.

A faster method is to use electronic mail where it is sent almost in an instance. For example a memo with 100 words will be sent in a fraction of a second. It is also simple to send to specific groups, various individuals, company-wide, and so on. Other types of data can also be sent with the mail message such as images, sound, and so on. It may also be possible to determine if a user has read the mail. The main advantages are:

- very much faster than traditional mail;
- easy to send to specific groups;
- more secure (in most cases);
- it is possible to determine if the recipient has read the message;
- it allows different types of data can be sent, for example, images, sound, and so on.

The main disadvantages are:

- it stops people using the telephone;
- it cannot be used as a legal document;
- electronic mail messages can be sent on the 'spur of the moment' and are then later regretted (sending by traditional methods normally allows for a re-think);
- it may be difficult to send to remote sites;
- not everyone reads their electronic mail on a regular basis.

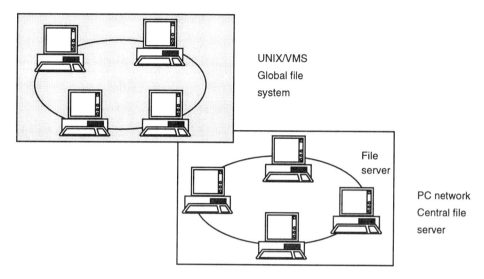

Figure 2.5 Sharing disk space with Unix/ VMS and PC network

2.2.6 Peer-to-peer communication

A major problem with computers is to make them communicate with a different computer type or with another that possibly uses a different operating system. A local network allows different types of computers running different operating systems to share information over the network. This is named peer-to-peer exchange and is illustrated in Figure 2.6.

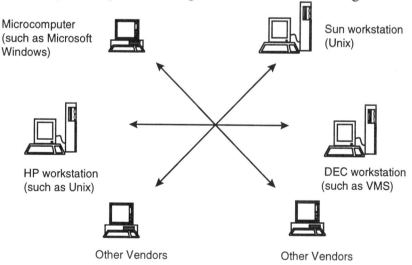

Figure 2.6 Peer-to-peer exchange over a network

2.2.7 Remote login

A major advantage with networks is that they allow users to remotely log into other computers. The computer being logged into must be running a multi-tasking operating system, such as Unix. Figure 2.7 shows an example of three devices (a workstation, an X-windows terminal and a PC) logging into a powerful workstation. This method allows many less powerful computers to be linked to a few powerful machines.

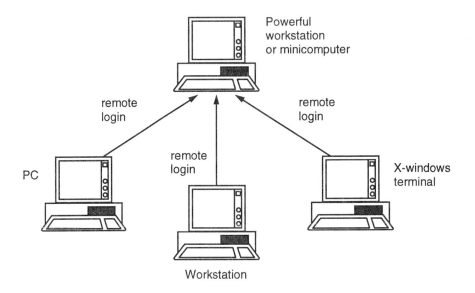

Figure 2.7   Remote login into other nodes

2.2.8 Disadvantages and potential pitfalls of networks

The main disadvantage of networks is that users become dependent upon them. For example if a network file server develops a fault then many users may not be able to run application program and get access to shared data. On many sites a back-up server can be switched into action when the main server fails. A fault on a network may also stop users from being able to access peripherals such as printers and plotters. To minimize this a network is normally segmented so that a failure in one part of it does not affect other parts.

Another major problem with networks is that their efficiency is very dependent on the skill of the system manager. A badly managed network may operate less efficiently than non-networked machines. Also, a badly run network may allow external users into it with little protection against them causing damage. Damage could also be causes by novices causing problems such as deleting important files.

The main disadvantages are summarized below:

- if network file server develops a fault then users may not be able to run application programs;
- a fault on the network can cause users to lose data (especially if they have not saved the current working file recently);
- if the network stops operating then it may not be possible to access various resources;
- users work-throughput becomes dependent upon network and the skill of the system manager;
- it is difficult to make the system secure from hackers, novices or industrial espionage (again this depends on the skill of the system manager);
- decisions on resource planning tend to become centralized, for example, what word processor is used, what printers are bought, and so on;
- networks that have grown with little thought can be inefficient in the long term;
- as traffic increases on a network the performance degrades unless it is designed properly;
- resources may be located too far away from some users;
- the larger the network becomes the more difficult it is to manage.

2.3 OSI MODEL

An important concept in understanding data communications is the OSI (open systems interconnection) model. It allows manufacturers of different systems to interconnect their equipment through standard interfaces. It also allows software and hardware to integrate well and be portable on differing systems. International Standards Organisation (ISO) developed the model and it is shown in Figure 2.8.

Data is passed from the top layer of the transmitter to the bottom and then up from the bottom layer to the top on the recipient. Each layer on the transmitter, though, communicates directly the recipients corresponding layer. This creates a virtual data flow between layers.

The top layer (the application layer) initially gets data from an application and appends it with data that the recipients application layer will read. This appended data passes to the next layer (the presentation layer). Again it appends its own data, and so on, down to the physical layer. The physical layer is then responsible for transmitting the data to the recipient. The data sent can be termed a data packet or data frame.

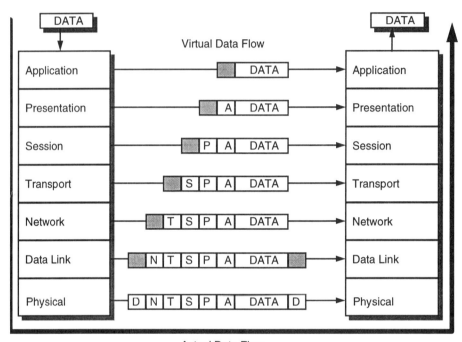

Figure 2.8   Seven-layer OSI model

Figure 2.9 shows the basic function of each of the layers. The physical link layer defines the electrical characteristics of the communications channel and the transmitted signals. This includes voltage levels, connector types, cabling, and so on.

The data link layer ensures that the transmitted bits are received in a reliable way. This includes adding bits to define the start and end of a data frame, adding extra error detection/ correction bits and ensuring that multiple nodes do not try to access a common communication channel at the same time.

The network layer routes data frames through a network. If data packets require to go out of a network then the transport layer routes it through interconnected networks. Its task may involve, for example, splitting up data for transmission and re-assembling it upon reception.

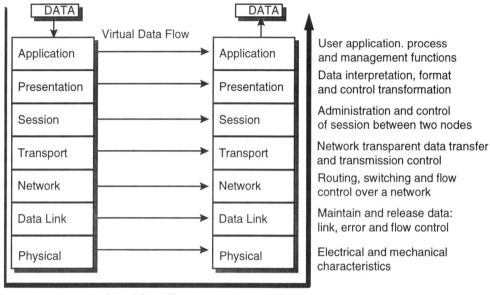

Figure 2.9 ISO open systems interconnection (OSI) model

The session layer provides an open communications path with the other system. It involves the setting up, maintaining and closing down a session. The communication channel and the internetworking of the data should be transparent to the session layer.

The presentation layer uses a set of translations that allows the data to be interpreted properly. It may have to carry out translations between two systems if they use different presentation standards such as different character sets or differing character codes. For example on a Unix system a text file the a new-line has one ASCII character, the carriage return. Whereas on a DOS-based system there are two, the line feed and carriage return characters. The presentation layer would convert from one computer system to another so that the data was displayed correctly, in this case by either adding or taking away a character. The presentation layer can also add data encryption for security purposes.

The application layer provides network services to application programs such as file transfer and electronic mail.

Figure 2.10 shows an example with two interconnected networks, Network A and Network B. Network A has four nodes N1, N2, N3 and N4, and Network B has nodes N5, N6, N7 and N8. If node N1 communicates with node N7 then a possible path would be via N2, N5 and N6. The data link layer ensures that the bits transmitted between nodes N1 and N2, nodes N2 and N5, and so on, are transmitted in a reliable way.

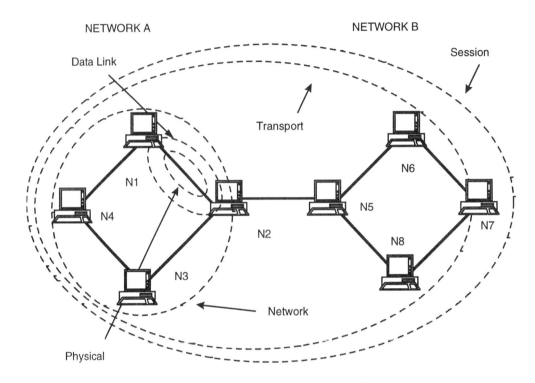

Figure 2.10   Scope of concern of OSI layers

The network layer would then be responsible for routing the data packets through Network A and through Network B. The transport layer routes the data through interconnection between the networks. In this case it would route data packets from N2 to N5. If other routes existed between N1 and N7 it may use another route.

2.4 COMMUNICATIONS STANDARDS AND THE OSI MODEL

The following sections look at practical examples of data communications and networks and how they fit into the layers of the ISO model. Unfortunately, most currently available technologies do not precisely align with the layers of this model. For example, RS-232 provides a standard for a physical layer but it also includes some data link layer functions, such as adding error detection and framing bits for the start and end of a packet.

Figure 2.11 shows the main technologies covered in this book. These are split into three basic sections: asynchronous data communication, local area networks (LANs) and wide area networks (WANs).

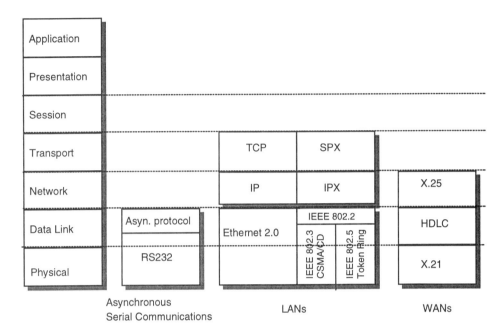

Figure 2.11 ISO open systems interconnection (OSI) model

The most popular types of LANs are Ethernet and token ring. Standards for Ethernet include Ethernet 2.0 and IEEE 802.3 (with IEEE 802.2). For token ring it is IBM token ring and IEEE 802.5 (with IEEE 802.2). Ethernet uses carrier sense multiple access/ collision detect (CMSA/CD) technology which is why the IEEE standard includes the name CSMA/CD.

One of the main standards for the interconnection of networks is the Transport Control Protocol (TCP)/ Internet Protocol (IP). The IP part routes data packets through a network and TCP routes packets between interconnected network. An equivalent standard TCP/IP used in some PC networks is SPX/IPX.

For digital connections to WANs the main standards are CCITT X.21, HDLC and CCITT X.25.

2.5 STANDARDS AGENCIES

There are six main international standards agencies that define standards for data communications system. These are the ISO (International Standards Organization), the CCITT (International Telegraph and Telephone Consultative Committee), the EIA (Electrical Industries Association), the ITU (International Telecommunications Union), ANSI (American National Standards Institute) and the IEEE (Institute of Electrical and Electronic Engineers).

The ISO and IEEE have defined standards for the connection of computers to local area networks and the CCITT have defined standards for the interconnection of national and international networks. The CCITT standards covered in this book split into three main sections, these are asynchronous communications (V.xx standards), PSDN connections (X.xxx standards) and ISDN (I.4xx standards). The main standards are given in Table 2.1.

The EIA have defined standards for the interconnection of computers using serial communications. The original standard was RS-232-C, this gives a maximum bit rate of 20 kbps over 20 m. It has since defined several other standards including RS-422 and RS-423 that provide a data rate of 10 Mbps.

2.6 NETWORK CABLE TYPES

The cable type used on a network depends on several parameters, including:

- the data bit rate;
- the reliability of the cable;
- the maximum length between nodes;
- the possibility of electrical hazards;
- power loss in the cables;
- tolerance to harsh conditions;
- expense and general availability of the cable;
- ease of connection and maintenance;
- ease of running cables, and so on.

The main types of cables used in networks are twisted-pair, coaxial and fibre optic, these are illustrated in Figure 2.12. Twisted-pair and coaxial cables transmit electric signals, whereas fibre optic cables transmit light pulses. Twisted-pair cables are not shielded and thus interfere with nearby cables. Public telephone lines generally use twisted-pair cables. In LANs they are generally used up to bit rates of 10 Mbps and with maximum lengths of 100 m.

Table 2.1   Typical standards

Standard	Equivalent ISO/CCITT	Description
EIA RS-232-C	CCITT V.28	Serial transmission up to 20 kps/ 20 m
EIA RS-422	CCITT V.11	Serial transmission up to 10 Mbps/ 1200 m
EIA RS-423	CCITT V.10	Serial transmission up to 300 kbps/ 1200 m
ANSI X3T9.5		LAN: Fibre Optic FDDI standard
IEEE 802.2	ISO 8802.2	LAN: IEEE standard for logical link control
IEEE 802.3	ISO 8802.3	LAN: IEEE standard for CSMA/CD
IEEE 802.4	ISO 8802.4	LAN: Token passing in a token ring network
IEEE 802.5	ISO 8802.5	LAN: Token ring topology
	CCITT X.21	WAN: Physical layer interface to a PSDN
HDLC	CCITT X.212/ 222	WAN: Data layer interfacing to a PSDN
	CCITT X.25	WAN: Network layer interfacing to a PSDN
	CCITT I430/1	ISDN: Physical layer interface to an ISDN
	CCITT I440/1	ISDN: Data layer interface to an ISDN
	CCITT I450/1	ISDN: Network layer interface to an ISDN

Coaxial cable has a grounded metal sheath around the signal conductor. This limits the amount of interference between cables and thus allows higher data rates. Typically they are used at bit rates of 100 Mbps for maximum lengths of 1 km.

The highest specification of the three cables is fibre optic. This type of cable allows extremely high bit rates over long distances. Fibre optic cables do not interfere with nearby cables and give greater security, give more protection from electrical damage from by external equipment, are more resistance to harsh environments and are safer in hazardous environments.

A typical bit rate for a LAN using fibre optic cables is 100 Mbps, in other applications this reach several gigabits/sec. The maximum length of the fibre optic cable depends on the transmitter and receiver electronics but a single length of 20 km is possible.

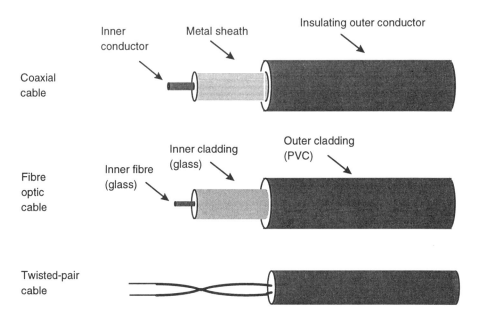

Figure 2.12   Typical network cable types

2.7 NETWORK TOPOLOGIES

Figure 2.13 shows the three basic topologies for a LAN, these are:

- a star network;
- a ring network;
- a bus network.

2.7.1 Star network

In a star topology, a central server switches data around the network. Data traffic between nodes and the server will be relatively low thus twisted-pair cable can connect the nodes to the server.

Another advantage of a star network is that a fault on one of the nodes will not affect the rest of the network. Typically, mainframe computers use a central server with terminals connected to it.

The main disadvantage of this topology is that the network is highly dependent upon the operation of the central server. If it where to slow down significantly then the network becomes slow. Also if it was to become un-operational then the complete network would shut down.

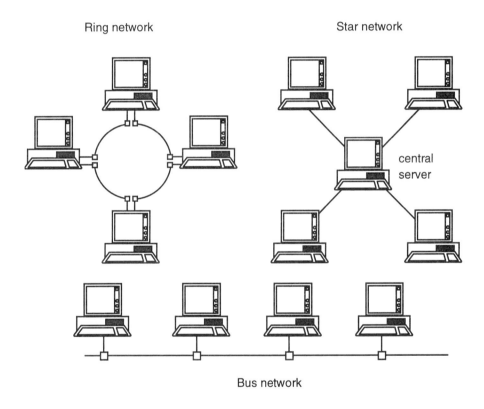

Figure 2.13 Network topologies

2.7.2 Ring network

In a ring network nodes connect to make a closed electronic loop. Each node communicates with its neighbour and thus data is passed from one node to the next until it reaches its destination. Normally to provide an orderly flow of data an electronic token passes around the ring, this is known as a token passing ring network, as shown in Figure 2.14.

A node which wishes to transmit data must wait for the token. One by one each node on the network reads the token, no matter if it contains data or not, and repeats it to the next node. A distributed control protocol determines the sequence in which nodes transmit.

In a manner similar to the star network each link between nodes is basically a point-to-point link. This allows almost any transmission medium to be used. Typically twisted-pair cables allow a bit rate of up to 10 Mbps, but coaxial and fibre optic cables are normally used for extra reliability and higher data rates. A typical ring network is IBM token ring.

A ring network gives high data rates but suffers from several problems. The most severe is that if one of the nodes goes down then the whole network may go down.

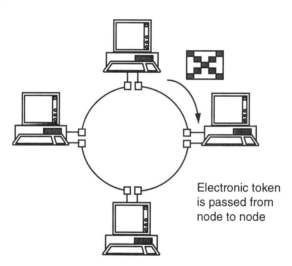

Electronic token is passed from node to node

Figure 2.14   Token passing ring network

2.7.3 Bus network

A bus network uses a multi-drop transmission medium, as shown in Figure 2.15. All nodes on the network share a common bus and all share communications. This allows only one device to communicate at a time. A distributed medium access protocol determines which station is to transmit. As with the ring network, data packets contain source and destination addresses. Each station monitors the bus and copies frames addressed to itself.

Twisted-pair cables gives data rates up to several Mbps. Coaxial and fibre optic cables give higher bit rates and longer transmission distances.

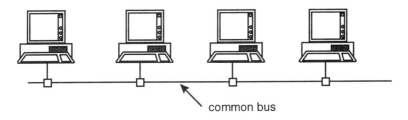

common bus

Figure 2.15   Bus topology

Two transmission methods are possible, these are baseband and broadband communications. Baseband communications uses a single carrier frequency. Whereas, broadband transmits data over a range of carrier frequencies and can thus transmit more data. Baseband communications is simpler and has a lower cost. The layout of a baseband cable is simple. Broadband communications are more complex to install and maintain but give a higher data rate.

A bus network is a good compromise over the other two topologies as it allows relatively high data rates. Also, if a node goes down then it does not affect the rest of the network. The disadvantage of this topology is that it requires a network protocol to detect when two nodes are transmitting at the same time. A typical bus network is Ethernet 2.0.

2.8 MANCHESTER CODING

Two of the network technologies covered in the following chapter, Ethernet and token ring, use Manchester coding for the bits sent on the transmission media. This coding has the advantage of embedding timing (clock) information within the transmitted bits. A positive edged pulse (low→high) represents a 1 and a negative edged pulse (high→low) a 0, as shown in Figure 2.16. Another advantage of this coding method is that the average voltage is always zero when used with equal positive and negative voltage levels.

Figure 2.17 is an example of transmitted bits using Manchester encoding. The receiver passes the received Manchester encoded bits through a low-pass filter. This extracts the lowest frequency in the received bit stream, that is, the clock frequency. With this clock the receiver can then determine the transmitted bit pattern.

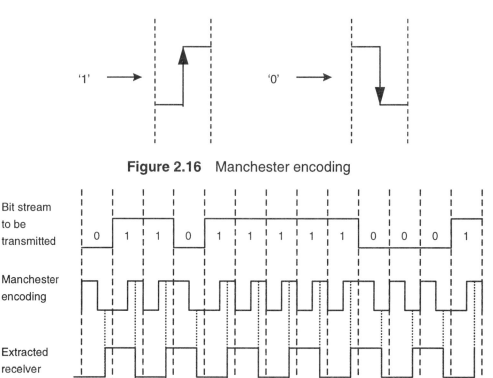

Figure 2.16   Manchester encoding

Figure 2.17   Example of Manchester coding

2.9 EXERCISE

2.1   For the following functions determine which of the layers of the ISO model that they fit into:

	P	D	N	T	S	P	A
Addition of bits for error control							
Addition of bits in order to route a packet through a network							

	P	D	N	T	S	P	A
Addition of bits to allow the data to be transported between interconnected networks							
Addition of bits to define the start and end of a data packet							
Setting up a login procedure from one computer to another							
Transmission of bits from the transmitter to the receiver							
Data encryption/ decryption							

2.2   Peer-to-peer communications involves:

    A   computers, possibly of differing types, to share information over a network

    B   communications between a computer and the network

    C   the communications between the computer connector and the network cable

    D   the interconnection of computers which have the same operating system

2.3   The main advantage of the ISO model is that:

    A   it defines how data communications equipment should connect

    B   it fits in well with most available network technologies

    C   it allows system developers to integrate hardware and software through standard interfaces

    D   it defines how cables and connectors should be used

2.4   The CCITT V. series standards relate primarily to which area of data communications:

A   Wide area networks
B   Asynchronous communications
C   Local area networks
D   Integrated services digital networks

2.5   The CCITT I. series standards relate primarily to which area of data communications:

A   Wide area networks
B   Asynchronous communications
C   Local area networks
D   Integrated services digital networks

2.6   The CCITT X. series standards relate primarily to which area of data communications:

A   Wide area networks
B   Asynchronous communications
C   Local area networks
D   Integrated services digital networks

2.7   Which organization developed the original RS-232-C specification:

A   EIA
B   ISO
C   IEEE
D   ITU

2.8   Which organization has developed much of the original standards for local area networks:

A   EIA
B   ISO
C   IEEE
D   ITU

2.10 TUTORIAL

2.9   Discuss the advantages of Manchester coding.

2.10   Show the voltage waveforms for the following Manchester coded bit patterns:

     (i)   `00000000`   (ii)   `11111111`   (iii)   `10101010`

2.11   Discuss the advantages and disadvantages of differing network topologies.

     For a known network, such as a university, college or company network complete questions Q2.12 to Q2.16.

2.12   For a known network discuss the advantages and disadvantages of the network. If possible, identify any strengths and any weaknesses.

2.13   For a known network list the main application programs on the file system. A sample table, for a PC network, is given in Table 2.2.

Table 2.2   Application programs on a known LAN

Type of application	Application name	Home directory
Word processor	Word for Windows Ver. 2.0	\WIN\WINWORD
Spreadsheet	Lotus 123 Ver. 2.0	\WIN\123
Presentation package	Microsoft Powerpoint Ver. 2.0	\WIN\PPT

2.14   For a known network, such as a company or university network, determine the type of networking cable it uses. If possible, determine why it was chosen.

2.15   For a known network, such as a company or university network, determine the network topology. If possible, draw a schematic of the network showing the connected nodes.

2.16   For a known network, list the networked peripherals. A sample table is given in Table 2.3.

Table 2.3   Peripherals connected to a known LAN

Peripheral	Number connected	Description
Laser printer	2	HP Laserjet 4
Line printer	1	Epson LX-400
Plotter	1	HP 7475A
Image scanner	1	HP Scanjet

3

LAN: Ethernet, token ring and FDDI

3.1 INTRODUCTION

A local area network (LAN) is a collection of data communications equipment connected by a common electronic connection within a relatively small area. Normally, LANs are owned and operated by their user. Standards for LAN have been agreed through the IEEE 802 and ANSI X3T9.5 committees. The IEEE standards have now been adopted by the ISO as international standards.

3.2 IEEE STANDARDS

Figure 3.1 shows how the IEEE standards for token ring and CSMA/CD fit into the OSI model. A CSMA/CD network is more commonly known as an Ethernet network.

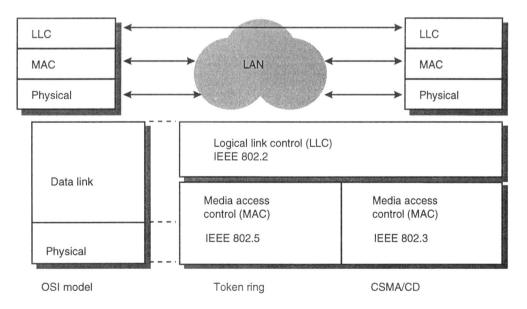

Figure 3.1 Standards for IEEE 802 LANs

The two layers of the IEEE standards correspond to the physical and data link layers of the OSI model. A token ring network uses IEEE 802.5 (ISO 8802.5) and a CSMA/CD network uses IEEE 802.3 (ISO 8802.3).

The IEEE 802.2 (ISO 8802.2) logical link control (LLC) layer conforms to the same specification for both types of network. A medium access control (MAC) unit allows many nodes to share a single communication channel. It also adds start and end frame delimiter, error detection bits, access control information and source and destination addresses.

3.3 LOGICAL LINK CONTROL

The LLC level provides some of the functions of a data link layer, these are:

- flow and error control – each data frame sent has a frame number. A control frame is sent from the destination node to the source node informing it that it has or has not received frames correctly;
- sequencing of data – large amounts of data are spliced and sent with frame numbers. The spliced data is then reassembled at the destination node.

The format of the LLC is similar to HDLC which will be covered in more detail in Chapter 6. Figure 3.2 shows the basic format of the LLC frame.

There are three basic types of frames: information, supervisory and unnumbered frames. An information frame contains data, a supervisory frame is used for acknowledgement and flow control, and an unnumbered frame is used for control purposes. The first 2 bits of the control field determines which type of frame it is. If they are 0X (where X is a don't care) then it is an information frame, a 10 specifies a supervisory frame and a 11 an unnumbered frame.

An information frame contains a send sequence number in the control field which ranges from 0 to 127. Each information frame has a consecutive number, N(S) (note that there is a roll-over from frame 127 to frame 0). The destination node acknowledges that it has received the frames by sending a supervisory frame. The function of the supervisory function is specified by the 2-bit S bit field. This can either be set to Receiver Ready (RR), Receiver Not Ready (RNR) or Reject (REJ). If a RNR function is set then the destination node acknowledges that all frames up to the number stored in the receive sequence number N(R) field were received correctly. A RNR function also acknowledges the frames up to the number N(R), but informs the source node that the destination node wishes to stop communicating. The REJ function specifies that frame N(R) has been rejected and all other frames up to N(R) are acknowledged.

Figure 3.3 shows an example of a source sending two information frames, numbered 1 and 2. A Receiver Ready control frame from the destination node acknowledges these frames. Next, the source sends three more information frames. The destination node then sends a rejection for information frame 4. The source then retransmits this frame and each frame after the reject frame number. Note that by rejecting frame number 4 that the destination node has acknowledged frame number 3, thus there is no need to retransmit it.

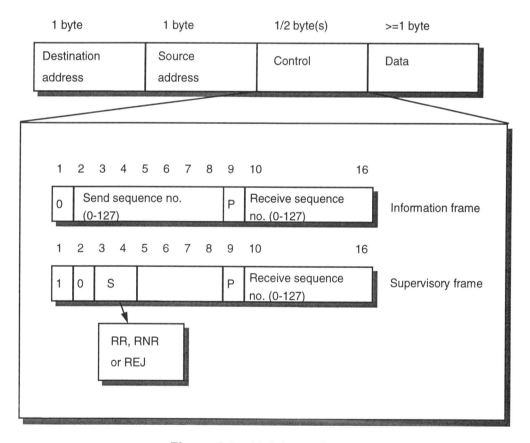

Figure 3.2   LLC frame format

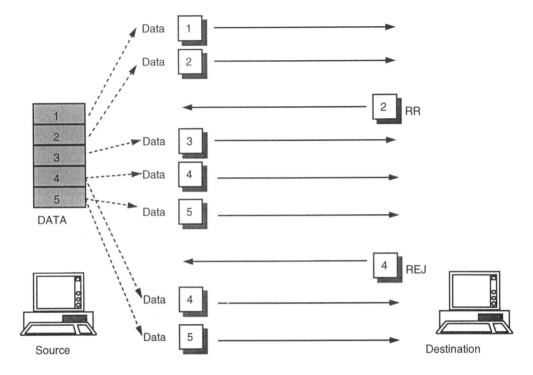

Figure 3.3   Example of acknowledgement of information frames

3.4 TOKEN RING

The IEEE 802.5 standard specifies the MAC layer for a token ring network with a bit rate of either 4 Mbps or 16 Mbps.

A token ring network circulates an electronic token around a closed electronic loop. Each node on the network reads the token and repeats it to the next node. The token circulates around the ring even when there is no data being transmitted.

A node wishing to transmit waits for a token. When it gets it, it fills a frame with data and adds the source and destination addresses and sends it to the next node. The data frame then circulates around the ring until it reaches the destination node. It then reads the data into its local memory area (or buffer) and marks an acknowledgement on the data frame. This then circulates back to the source (or originating) node. When it receives the frame it tests it to determine if it contains an acknowledgement. If it does then the source nodes knows that the data frame was received correctly, else the node is not responding. If the source node has finished transmitting data then it transmits a new token, which can be used by other nodes on the ring.

A distributed control protocol determines the sequence in which nodes transmit. This gives each node equal access to the ring.

Token ring is well suited to networks which have large amounts of traffic and also works well with most traffic loadings. It is not suited to large networks or networks with physically remote stations.

Its main advantage is that it copes better with high traffic rates than Ethernet (to be covered next), but requires a great deal of maintenance especially when faults occur or when new equipment is added to or removed from the network.

3.4.1 Token ring - media access control (MAC)

Token passing allows nodes controlled access to the ring. Figure 3.4 shows the token format for the IEEE 802.5 specification. There are two main types of frames: a control token and a data frame. A control token contains only a start and end delimiter, and an access control (AC) field. A data frame has start and end delimiters (SD/ED), an access control field, a frame control field (CF), a destination address (DA), a source address (SA), frame check sequence (FCS), data and a frame status field (FS).

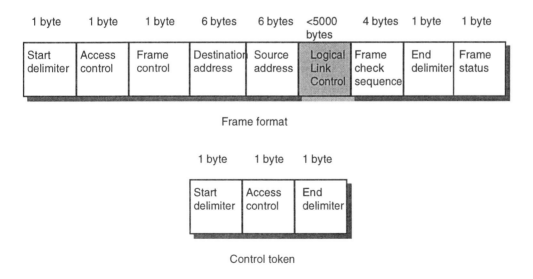

Figure 3.4   IEEE 802.5 frame format

The access control and frame control fields contain information necessary for managing access to the ring. This includes priority reservation,
priority information and information on whether the data is user data or
control information. It also contains an express indicator which informs
networked nodes that an individual node requires immediate action from
the network management node.

The destination and source addresses are either 2 or 6 bytes (or octets)
in length. Logical link control information is variable length and is shown
in Figure 3.2. It can either contain user data or network control information.

The frame check sequence (FCS) is a 32-bit cyclic redundancy check
(CRC) and the frame control field is used to indicate whether a destination node has read the data in the token.

Start and end delimiter

The delimiter is a special bit sequence which defines the start and end of a
frame and cannot occur anywhere within the frame. This is achieved by
changing the Manchester coding scheme in the start and end delimiters.
Two of the bits within the delimiters are set to either a high level (H) or a
low level (L). These bits disobey the standard coding as there is no
change in level, that is, from a high to a low or a low to a high. When the
receiver detects this volition and the other standard coded bits in the received bit pattern then it knows that the bits that follow are a valid frame.
The coding is as follows:

- if the preceding bit is a 1 then the start delimiter is HL0HL000, else
- if the preceding bit is a 0 then the start delimiter is LH0LH000. These
  are shown in Figure 3.5.

The end delimiter is similar to the start delimiter, but 0's are replaced by
1's. An error detection bit (E) and a last packet indicator bit (I) are added.

If the bit preceding the end delimiter is a 1 then the end delimiter is
HL1HL1IE. If it is a 0 then it is LH1LH1IE. The E bit is used for error
detection and is initially set by the originator to a 0. If any of the nodes on
the ring detects an error it is set to a 1. This indicates to the originator that
the frame has developed an error as it was sent. The I bit determines if
the data being sent in a frame is the last in a series of data frames. If it is a
0 then it is the last, else it is an intermediate frame.

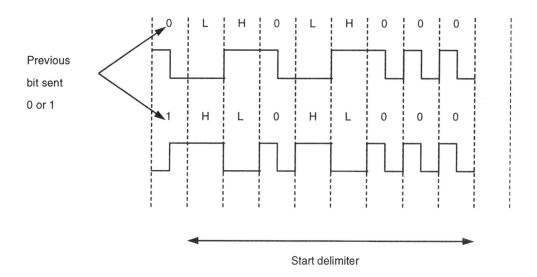

Figure 3.5   Start delimiter

Access control field

The access control field controls the access of nodes on the ring. It takes the form of PPPTMRRR, where:

PPP – indicates the priority of the token. This indicates which type of token the destination node can transmit;

T   – is the token bit and is used to discriminate between a control token and a data token;

M   – is the monitor bit and is used by an active ring monitor node to stop tokens from a circulating around a network continuously;

RRR – are the reservation bits and allow nodes with a high-priority to request the next token.

Frame control field

The frame control field contains controls information for the MAC layer. It takes the form of FFDDDDDD, where:

FF   – indicate whether the frame is a data frame. If it is not then the DDDDD bits control the operation of the token ring MAC protocol;

DDDDDD – control the operation of the token ring MAC protocol.

Source and destination addresses

The source and destination addresses can either be 2 or 6 bytes (that is, 16 or 48 bits) in length. This size must be the same for all nodes on a ring. The first bit specifies the type of address. If it is a 0 then the address is an individual node address, else it is a group address. An individual node address is used to transmit to a single node, whereas a group address transmits to all nodes with the same group address. The source address will always by an individual address as it indicates the node which originated the token. A special destination address of all 1's is used to transmit to all nodes on a ring.

Frame check sequence

The frame check sequence (or FCS) is an error detection scheme. It is used to determine transmission errors and is often referred to as a cyclic redundancy check (CRC) or simply as checksum.

Frame status

The frame status field contains information on how a frame has been operated on as it circulates round the ring. It takes the form of ACXXACXX, where:

A –   indicates if the destination address has been recognized. It is initially set to a 0 by the source node and is set to a 1 when the destination reads the data. If the source node detects that this bit has not been set then it knows that the destination is either not present on the network or is not responding;

C –   indicates that a destination node has copied a frame into its memory. This bit is also initially set to a 0 by the source node. When the destination node reads the data from the frame it is set to a 1. By testing this bit and the A bit the source node could determine if the destination node is active but not reading data from the frame.

3.4.2 Token ring maintenance

A token ring system requires considerable maintenance and it must perform the following:

• ring initialization – when the network is started, or after the ring has been broken, it must be re-initialized. A co-operative decentralized algorithm sorts out which nodes start a new token, which goes next, and so on;

- adding to the ring – if a new node is to be physically connected to the ring then the network must be shut down and re-initialized;
- deletion from ring – a node can disconnects itself from the ring by joining together its predecessor and successor. Again, the network may have to be shut down and re-initialized;
- fault management – typical token ring errors are when two nodes think its their turn to transmit or when the ring is broken as no node thinks that it is their turn.

3.4.3 Token ring multi-station access units (MAU)

The problems of connecting and deleting nodes to/from a ring network can be reduced by using a multi-station access unit (MAU). Figure 3.6 shows two three-way MAUs connected to produce a six-node network. Normally, a MAU allows nodes to be switched in and out of a network using a changeover switch or by automatic electronic switching. This has the advantage of not shutting down the network when nodes are added, deleted or when they develop faults.

If the changeover switches in Figure 3.6 are in the down position then the node is bypassed, else if they are in the up position then the node connects to the ring.

A single coaxial (or twisted-pair) cable connects one concentrator to another and two coaxial (or twisted-pair) cables connect a node to the concentrator (for the in and the out ports).

The IBM 8228 is a typical passive MAU. It can operate at 4 Mbps or 16 Mbps and has 10 connection ports, that is, eight passive node ports, and a ring in (RI) and a ring out (RO) connection. The maximum distance between MAUs units is typically 650 m (at 4 Mbps) and 325 m (at 16 Mbps).

Most MAUs either have 2-, 4- or 8-ports. Figure 3.7 shows a 32-node token ring network using four 8-port MAUs. Typical connectors are RJ-45 and IBM type A connectors. The ring cable is normally either twisted-pair, fibre optic or coaxial cable. MAU units are intelligent devices and can detect faults on the cables supplying nodes and can then isolate these from the rest of the ring. Most MAUs are passive device in that they do not require a power supply. If there are large distances between nodes then an active unit is normally used.

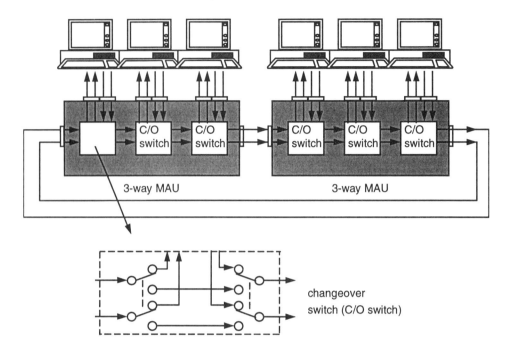

Figure 3.6   Six-node token ring network with two MAUs

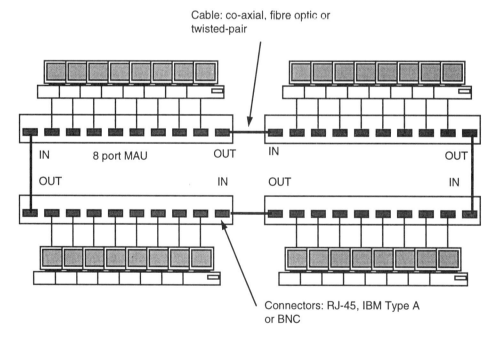

Figure 3.7   32 node token ring network with 4 MAUs

3.5 ETHERNET

The Xerox Corporation, in conjunction with DEC and Intel, initially developed Ethernet. International standards have since been developed by the IEEE 802 committee. It uses a bus network topology where all nodes share a common bus and only one node can communicate at a time. Data frames contain source and destination addresses and each node monitors the bus and copies any frames addressed to itself.

Ethernet uses carrier sense, multiple access with collision detection (CSMA/CD). On a CSMA/CD network, nodes monitor the bus (or Ether) to determine if it is busy. A node wishing to send data waits for an idle condition then transmits its message. A collision can thus occur when two nodes transmit at the same time, thus nodes must monitor the cable when they transmit. When this happens both nodes stop transmitting frames and transmit a jamming signal. This informs all nodes on the network that a collision has occurred. Each of the nodes then waits a random period of time before attempting a re-transmission. As each node has a random delay time then there can be a prioritization of the nodes on the network. Nodes thus contend for the network and are not guaranteed access to it. Collisions generally slow down the network. Each node on the network must be able detect collisions and be capable of transmitting and receiving simultaneously.

Ethernet 2.0 and IEEE 802.3 are the most popular CSMA/CD standards and uses baseband communications. The transmission rate of Ethernet is 10 Mbps over a coaxial line, although some networks allow bit rates of 100 Mbps. Its advantages are that it is simple to use and easy to install. Unfortunately, its performance degrades with heavy traffic.

3.5.1 Ethernet - media access control (MAC) layer

Figure 3.8 gives the IEEE 802.3 frame format. It contains 2 or 6 bytes for the source and destination addresses (16 or 48 bits each), 4 bytes for the CRC (32 bits), 2 bytes for the LLC length (16 bits). The LLC part may be up to 1500 bytes long. The preamble and delay components define the start and end of the frame. The initial preamble and start delimiter are, in total, 8 bytes long and the delay component is a minimum of 96 bytes long.

Preamble and delay

A 7-byte preamble precedes the Ethernet frame. Each byte has a fixed binary pattern of 10101010 and each node on the network uses it to synchronize their clocks and transmission timings. It also informs nodes that a frame is to be sent and for them to check the destination address in the frame.

The end of the frame is a 96-byte delay period which provides the minimum delay between two frames. This slot time delay allows for the worst-case network propagation delay.

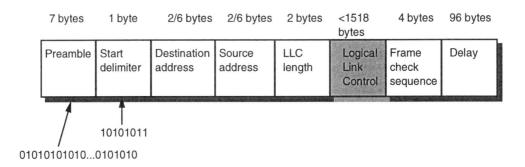

7 bytes	1 byte	2/6 bytes	2/6 bytes	2 bytes	<1518 bytes	4 bytes	96 bytes
Preamble	Start delimiter	Destination address	Source address	LLC length	Logical Link Control	Frame check sequence	Delay

10101011

01010101010...0101010

Figure 3.8   IEEE 802.3 frame format

Start delimiter

The start delimiter is a single byte (or octet) of 10101011. It follows the preamble and identifies that there is a valid frame being transmitted.

Source and destination address

The source address is either the 16- or 48-bit MAC address of the sending node and the destination address is the 16- or 48-bit Ethernet address of the destination node. All stations on the network must either have a 16- or a 48-bit address.

Each Ethernet and token ring devices have unique MAC address, this is normally defined in as hexadecimal digits, such as 4C – 31 – 22 – 10 – F1 – 32 or 4C31 : 2210: F132. A 48-bit address field allows 2^{48} different address (or approximately 280 000 000 000 000 addresses).

LLC length field

This field defines whether the frame contains information or it can be used to define the number of bytes in the logical link field.

Logical link field

The logical link field can contain up to 1518 bytes of information and has a minimum of 46 bytes, its format is given in Figure 3.2. If the information is greater than the upper limit then multiple frames are sent. Also, if the field is less than the lower limit then it is padded with extra redundant bits.

Frame check sequence

The frame check sequence (or FCS) is an error detection scheme. It is used to determine transmission errors and is often referred to as a cyclic redundancy check (CRC) or simply as checksum.

3.5.2 Hardware

Ethernet requires a minimal amount of hardware. Coaxial cables with an impedance 50 Ω normally connect nodes onto the common ether. Each node has transmission and reception hardware to control access to the cable and also to monitor network traffic. A node has frame building hardware to match the bits to the required Ethernet format. The transmission/ reception hardware is called a transceiver (short for *trans*mitter/re*ceiver*) and a controller builds up and strips down the frame. These are shown in Figure 3.9.

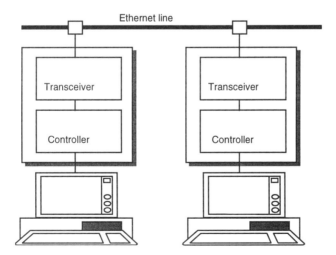

Figure 3.9 Ethernet transceiver and controller

Transceiver

Baseband Ethernet transmits onto a single ether. When the ether is not busy there is a voltage of +0.7 V on the line. This provides a carrier sense signal for all nodes on the network, it is also known as the heartbeat.

When a node requires to transmit a message it listens for a quiet period. If two or more transmitters transmit at the same time then a collision results. Each of the transceivers then broadcast a 'jam' signal. Upon reception of this signal, each node involved in the collision waits for a random period of time (ranging from 10 to 90 ms) before attempting to transmit again. Each transceiver on a network also awaits a retransmission.

When transmitting, a transceiver unit transmits the preamble of consecutive 1's and 0's. The coding used is a Manchester code which represents a 0 as a high to a low voltage transition and a 1 as a low to high voltage transition. A low voltage is −0.7 V and a high is +0.7 V. Thus when the preamble is transmitted the voltage will change between +0.7 and −0.7 V, this is illustrated in Figure 3.10. If after the transmission of the preamble no collisions are detected then the rest of the frame is sent.

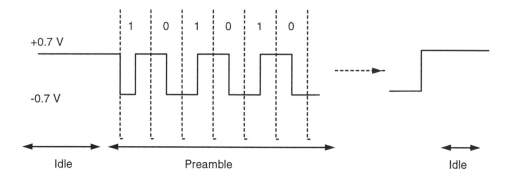

Figure 3.10 Ethernet digital signal

A transceiver connects a node to the coaxial cable. Its main functions are to:

- receive and transmit bits;
- broadcast a jam signal;
- test for reception;
- sense a collision;
- test for a heartbeat;
- transmit preamble;

- transmit delay signal.

Transceiver hardware normally detects collisions by monitoring the DC (or average) voltage on the ether.

Controller
The controller builds up and strips down the frame and it also adds the source and destination address. Its main functions are to:

- convert bits from serial to parallel and parallel to serial;
- store data in frames for transmission and reception;
- encode and decode broadcast signals;
- recognize addresses;
- detect errors and collisions;
- add and parse preamble;
- provide carrier sense and deference;
- filter collision fragments;
- generate and verify frame check sequences;
- filter alignment errors and overruns;
- limit data rate to prevent overruns;
- manage receive and transmit links;
- build up and disassemble frames;
- request re-transmission.

The controller disassembles received frames and passes the information to the receiving nodes operating system.

3.5.3 Ethernet limitations

There are various limitations on an Ethernet CSMA/CD system. These limitations relate to the maximum signal propagation times and as a consequence, maximum cable lengths. The limitations are also a function of the clock period.

Length of segments
Twisted-pair and coaxial cables have a characteristic impedance. If a cable is terminated with its characteristic impedance then there will be no

loss of power and no reflections at terminations. A coaxial cable must be terminated by its characteristic impedance (50 Ω, normally). The Ethernet connection may consist of many spliced coaxial sections. One or many sections constitute a cable segment, which is a stand-alone network. A segment must not exceed 500 metres. This is shown in Figure 3.11.

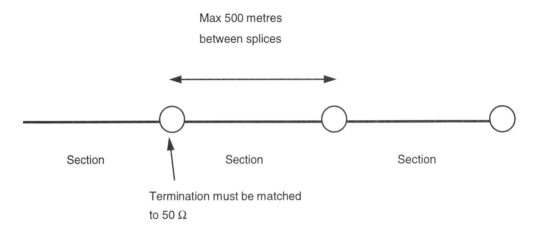

Figure 3.11 Connection of sections

Repeater lengths

A repeater is added between segments to boost the signal. A maximum of two repeaters can be inserted into the path between two nodes. The maximum of distance between two nodes connected via repeaters is 1 500 metres, this is illustrated in Figure 3.12.

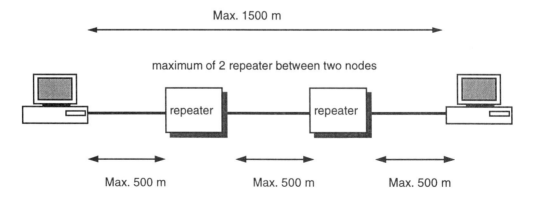

Figure 3.12 Maximum number of repeaters between two nodes

Maximum links

The maximum length of a point-to-point coaxial link is 1 500 metres, as shown in Figure 3.13. A long run such as this is typically used as a link between two remote sites within a single building.

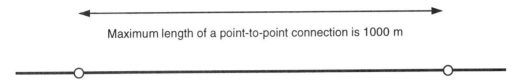

Maximum length of a point-to-point connection is 1000 m

Figure 3.13 Maximum length of a point-to-point connection

Distance between transceivers

Transceivers should not be placed closer than 2.5 metres. Additionally, each segment should not have more than 100 transceiver units, as illustrated in Figure 3.14. Transceivers which are placed too close to each other can cause transmission interference and also an increased risk of collision.

Each node transceiver lowers network resistance and dissipates the transmission signal. A sufficient number of transceivers reduces the electrical characteristic of the network below the specified operation threshold.

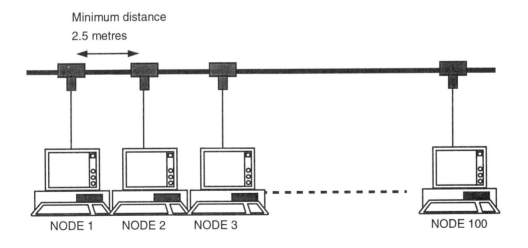

Maximum number of nodes on one segment is 100

Figure 3.14 Connection of sections

3.5.4 Ethernet types

There are four main types of Ethernet:

- Standard, or thick-wire, Ethernet (10BASE5);
- Thinnet, or thin-wire Ethernet, or Cheapernet (10BASE2);
- Twisted-pair Ethernet (10BASE-T);
- Optical fibre Ethernet (10BASEF).

The connection of the first three types to a Ethernet backbone are shown in Figure 3.15.

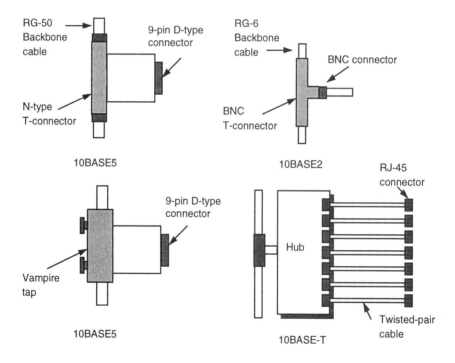

Figure 3.15   Ethernet connections

The standard Ethernet, 10BASE5, uses a high specification cable (RG-50) and N-type T-connectors to connect the transceiver to the backbone. A node connects to the transceiver using a 9-pin D-type connector. A vampire connector can be used to clamp the transceiver to the backbone cable.

Cheapernet uses a lower specification cable (it has a lower inner conductor diameter). The cable connector required is also of a lower specification, that is, BNC rather than N-type connectors. In standard Ethernet the transceiver unit is connected directly onto the backbone tap. On a Cheapernet network the transceiver is integrated into the node.

The cheapest of the three is 10BASE-T followed by 10BASE2. These have inferior electrical characteristics compared with 10BASE5. The basic specification for the types is given in Table 3.1.

Table 3.1 Ethernet network parameters

Parameter	*10BASE5*	*10BASE2*	*10BASE-T*
Common name	Standard or thick-wire Ethernet	Thinnet, thin-wire Ethernet or Cheapernet	Twisted-pair Ethernet
Data rate	10 Mbps	10 Mbps	10 Mbps
Maximum segment length	500 m	200 m	100 m
Maximum nodes on a segment	100	30	
Maximum nodes per network	1024	1024	
Minimum node spacing	2.5 m	0.5 m	
Maximum interconnected segments	3	3	
Location of transceiver electronics	located at the cable connection	integrated within the node	in a hub
Typical cable type	RG-50 (0.5" diameter)	RG-6 (0.25" diameter)	22 AWG twisted-pair
Connectors	N-type	BNC	RJ-45/ Telco
Cable impedance	50 Ω	50 Ω	
Transceiver cable connector	15 pin D-type	BNC	

3.5.5 Fast Ethernet

New standards relating to 100 Mbps Ethernet are now becoming popular. There are three main standards, these are 100BASE-TX (twisted-pair), 100BASE-T4 (twisted-pair) and 100BASE-FX (fibre optic cable). The 100BASE-T standards are compatible with 10BASE-T networks and allow both 10 Mbps and 100 Mbps bit rates on the line. This makes it easy to up-grade a network simply by add dual speed interface adapters. Nodes with the 100 Mbps capabilities can communicate at 100 Mbps, while they can also communicate with other, slower nodes, at 10 Mbps. Refer to Appendix D for more information.

3.5.6 Twisted-pair hubs

Twisted-pair Ethernet (10BASE-T) nodes normally connect to the backbone with a star connection using a hub, as illustrated in Figure 3.16.

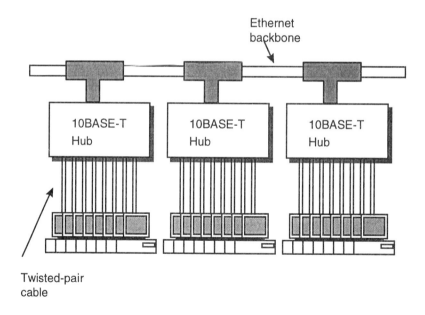

Figure 3.16   10BASE-T connection

3.5.7 Interconnection of LANs

If the performance of a network degenerated because of the amount of traffic on the medium then the network can be segmented by splitting it into two or more parts, each of which connect via a network bridge. The network bridge only routes frames from one segment to another if they need to go to the other segment.

Thus an Ethernet bridge, over time, learns the topology of the network. It does this by initially sending all frames between the two networks on either side of itself and the source and destination addresses are placed in its routing table. In time, this allows the bridge to decide whether to pass a particular frame on, or not. This is because its routing table contains which side of the bridge the frame destination is located.

In Figure 3.17 the bridge would contain a routing table with A, B, C, D, E on the top and F, G, H, I, J on the bottom. The traffic is thereby split and the load on each side reduced, thus enhancing performance.

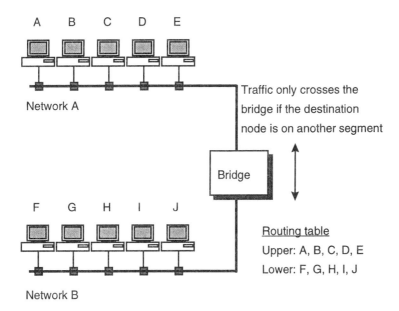

Figure 3.17   Ethernet bridge

A spanning-tree bridge allows multiple network segments to be inter-connected. If more than one path exists between individual segments then the bridge finds alternate routes. This is useful in routing frames away from heavy traffic routes or around a faulty route. Conventional bridges can cause frames to loop around forever. Spanning-tree bridges have built-in intelligence and can communicate with other bridges. This allows them to build up a picture of the complete network and thus make decisions on where frames are routed.

A network with two bridges is illustrated in Figure 3.18. In this case a tree spanning bridge could be inserted between network A and C so that traffic between the two networks does not affect network B.

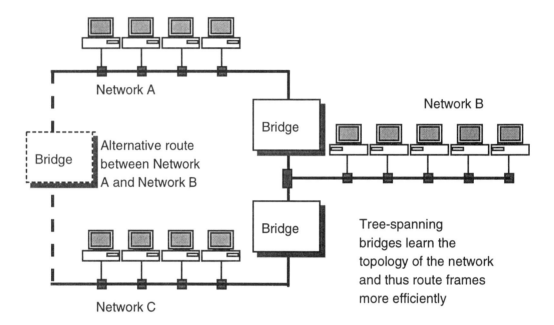

Figure 3.18   Network with alternative routes

3.6 FIBRE DISTRIBUTED DATA INTERCHANGE (FDDI)

Fibre optic cables allow a high specification over copper cables. The most common fibre optic network standard is FDDI. This operates at a 100 Mbps with two concentric token rings, as illustrated in Figure 3.19. Each node connected to the FDDI highway can be a normal node or a bridge to a conventional local area network, such as Ethernet or token ring. The maximum circumference of the ring is 100 km, with a maximum 2 km between nodes. A maximum of 1000 nodes can be connected to the network (500 per ring).

As there are two rings it is possible to use each of them for separate traffic streams. This effectively doubles the data carrying capacity of FDDI (to 200 Mbps). However, if the normal traffic is more than the stated carrying capacity or if one ring fails then its performance will degrade. The main advantage of the concentric rings is that the ring copes with fault conditions using loop-back.

FDDI uses a token passing medium access method. There are two types of token: a restricted and an unrestricted token. Under normal operating conditions an unrestricted token is used. When a node captures an unrestricted token it transmits frames for a period of time made up of a fixed part (T_f) and a variable time (T_v). This variable time depends on the traffic on the ring. With light traffic a node may keep the token much longer than when there is heavy traffic.

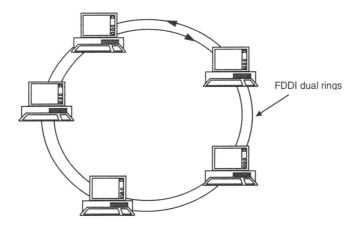

Figure 3.19   FDDI network

If a node wishes to enter an extended data interchange with another node, it captures the unrestricted token and changes to a restricted token. This allows the two nodes to communicate for an extended period, during which other nodes may only use the ring for the fixed period of time T_f. Once the data exchange is complete, or the extended time is over, the token is changed back to an unrestricted type.

FDDI uses a timed token-passing protocol to transmit data because a station can only hold the token up to a specified amount of time. Therefore, there is a limit to the amount of data that a node can transmit on any given opportunity.

A sending node is responsible for issuing a new token after transmitting packets. The node directly downstream from a sending node has the new opportunity to capture the token. These features and the timed token ensures that the ring's capacity is divided almost equally among the nodes on the ring.

3.6.1 Applications of FDDI networks

Typical applications of FDDI networks are:

- as a sub-network connecting high-speed computers and their peripheral devices (such as storage units);
- as a network connecting terminals, PCs and workstations where an application program requires high-speed transfers of large amounts of data (such as computer-aided design – CAD). Maximum data traffic for an FDDI network is at least 10 times greater than for standard Ethernet and token ring networks. As it is a token passing network it is less susceptible to heavy traffic loads than Ethernet;
- any applications requiring security and/or a high degree of fault tolerance. Fibre optic cables are generally more reliable and are difficult to tap-into without it being detected;
- as a backbone network in an internetwork connection.

3.6.2 FDDI backbone network

The backbone of a network is important as many users on the network depend on it. If the traffic is too heavy, or if it develops a fault, then it affects the performance of the whole network. An FFDI backbone helps with these problems by giving a high bit-rate and, normally, increasing the reliability of the backbone. An FDDI backbone is shown in Figure 3.20.

There are four types of nodes which can attach to an FDDI network:

- dual attachment stations (DAS);
- single attachment stations (SAS);
- dual attachment concentrators (DAC);
- single attachment concentrators (SAC).

Figure 3.21 shows an FDDI network configuration that includes all of these types of nodes.

An SAS connects to the FDDI rings through a concentrator, it is thus easy to add, delete or change the location of SASs. The concentrator automatically bi-passes disconnected nodes.

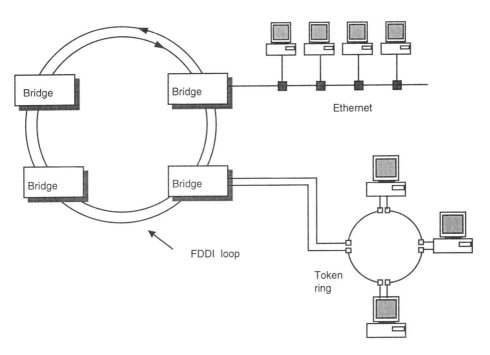

Figure 3.20   FDDI backbone network

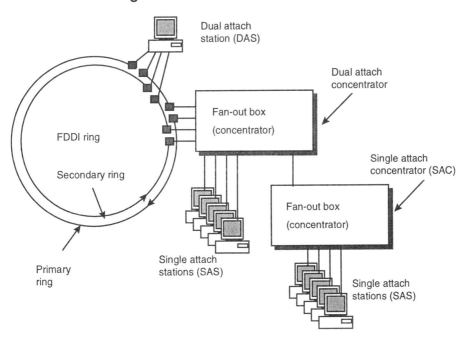

Figure 3.21   FDDI network configuration

Each DAS and DAC requires four fibres to connect it to the network:

Primary In, Primary Out, Secondary In and Secondary Out. The connection of a SAS only requires two fibres. Normally Slave In and Slave Out on the SAS connects to the Master Out and Master In on the concentrator unit, as shown in Figure 3.22.

FDDI stations attach to the ring using a Media Interface Connector (MIC). An MIC receptacle connects to the stations and an MIC plug on the network end of the connection. A dual attachment station has two MIC receptacles. One provides Primary Ring In and Secondary Ring Out, the other has Primary Ring Out and Secondary Ring In, as illustrated in Figure 3.23.

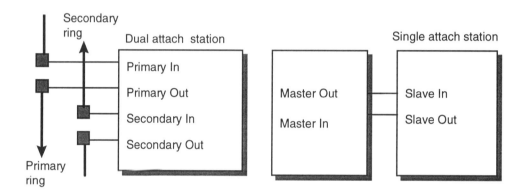

Figure 3.22   Connection of a DAS and a SAS

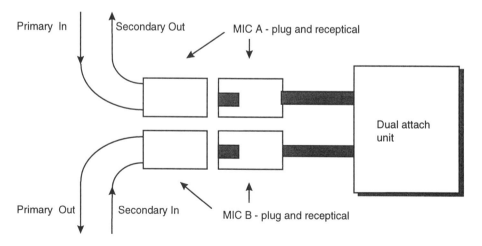

Figure 3.23   Connection of dual attach units

3.7 EXERCISE

3.1   Which of the following connects LAN to another of the same type:

A   a network backbone
B   a network trunk route
C   a network bridge
D   a network gateway

3.2   Which of the following connects a LAN to a WAN:

A   a network backbone
B   a network trunk route
C   a network bridge
D   a network gateway

3.3   The main standards relating to Ethernet networks are:

A   IEEE 802.2 and IEEE 802.3
B   IEEE 802.3 and IEEE 802.4
C   ANSI X3T9.5 and IEEE 802.5
D   EIA RS-422 and IEEE 802.3

3.4   Which layer in the Ethernet standard communicates with the OSI network layer:

A   the MAC layer
B   the LLC layer
C   the Physical layer
D   the Protocol layer

3.5   Standard, or thick-wire, Ethernet is also known as:

A   10BASE2
B   10BASE5
C   10BASE-T
D   10BASEF

3.6   Twisted-pair Ethernet is also known as:

| |
|✓|
| |
| |
| |
| |

A   10BASE2
B   10BASE5
C   10BASE-T
D   10BASEF

3.7   Fibre optic Ethernet is also known as:

A   10BASE2
B   10BASE5
C   10BASE-T
D   10BASEF

3.8   Which type of connector does standard Ethernet use when connecting to the network backbone:

A   N-type
B   BNC
C   RJ-45
D   Optical connector

3.9   Which type of connector does twisted-pair Ethernet use when connecting to a network hub:

A   N-type
B   BNC
C   RJ-45
D   Optical connector

3.10   Which type of connector does Cheapernet, or thin-wire Ethernet, use when connecting to the network backbone:

A   N-type
B   BNC
C   RJ-45
D   Optical connector

3.11 The main difference between standard Ethernet and Cheapernet are:

    A   that they have different bit rates
    B   the specifications for the cables and connectors
    C   that they have different frame formats
    D   that they use a different network topology

3.12 What is the function of a repeater in an Ethernet network:

    A   it increases the bit rate
    B   it isolates network segments
    C   it prevents collisions
    D   it boosts the electrical signal

3.13 What is the main function of a network bridge:

    A   it boosts the electrical signal
    B   it increases the bit rate
    C   it routes frames from one network to another
    D   it isolates nodes which have faults

3.14 How does a network bridge know how to route frames between networks:

    A   it is configured by the network manager
    B   it is pre-programmed when it is installed
    C   it learns the topology over time
    D   it routes all frames between networks

3.15 What is the main advantage of tree-spanning network bridges:

    A   they allow hubs to be connected
    B   they can communicate with other bridges to built up a complete picture of interconnected networks
    C   they increase the bit rate
    D   they store frames over a period of time

3.16   For a 4 Mbps bit rate determine the time period of each bit transmitted:

A   0.25 μs
B   2.5 μs
C   25 μs
D   4 μs

3.17   For a 10 Mbps bit rate determine the time period of each bit transmitted:

A   0.01 μs
B   0.1 μs
C   1 μs
D   10 μs

3.18   For a 16 Mbps bit rate determine the time period of each bit transmitted:

A   0.0625 μs
B   0.625 μs
C   6.25 μs
D   16 μs

3.19   How many fibres connect to a FDDI dual attach station:

A   1
B   2
C   3
D   4

3.8 TUTORIAL

3.20   For a known network, such as a university or company network, write a report containing some of the following:

- basic network topology (such as, ring, star, bus, and so on);
- network technology (such as, Ethernet, token ring, and so on);
- node connectors (such as, RJ-45, BNC, N-type, and so on);
- backbone cabling (such as, fibre optic, coaxial, and so on);
- bit rates (such as,. 4 Mbps, 10 Mbps, 16 Mbps, 100 Mbps, and so on);
- location of bridges and hubs.

3.21 Discuss how a destination node uses the LLC layer to indicate that frames have been received in error.

3.22 Discuss how token ring defines the start and end delimiters.

3.23 Discuss the main problems of a token ring network and describe some methods which can be used to overcome them.

3.24 Discuss the main advantages of MAUs in a token ring network.

3.25 Discuss the main reasons for the preamble in an Ethernet frame.

3.26 Discuss the limitations of the different types of Ethernet.

3.27 Determine the number of unique addresses that can be used for a 2-byte network address field. Compare this number with the number of addresses with a 6-byte address field.

3.28 A node has a binary network address of 0011 1111 0101 1111 1000 1000 0101 0000 0000 1000 0111 1111. Determine its hexadecimal address, that is, the address in the form XXXX : XXXX : XXXX. Table 3.2 shows the conversion of binary digits into hexadecimal.

For example to convert the binary number 01110101110000 into hexadecimal the following is conducted:

Binary	0111	0101	1100	0000
Hex	7	5	C	0

Table 3.2   Decimal and hexadecimal conversions

Decimal	Binary	Hex
0	0000	0
1	0001	1
2	0010	2
3	0011	3
4	0100	4
5	0101	5
6	0110	6
7	0111	7
8	1000	8
9	1001	9
10	1010	A
11	1011	B
12	1100	C
13	1101	D
14	1110	E
15	1111	F

3.29   Discuss the token passing technique used in FDDI.

4

LAN performance

4.1 INTRODUCTION

Many factors influence the choice of LAN technology. The most important are its performance, its cost of installation and maintenance, its compatibility with other equipment, long term availability of hardware and software, and its general reliability. A major concern is the behaviour of the network under heavy loads. The IEEE 802 committee sponsored Bell Labs to analyze the performance of token ring and CSMA/CD networks. Their general conclusions were:

- the smaller the mean frame length, the greater the difference in maximum mean throughput rate between token ring and CSMA/CD.
- token ring is the least sensitive to work load;
- CSMA/CD offers the shortest delay under light loads, while it is most sensitive under heavy load to the work load.

The following sections discuss possible ways of improving the performance of an Ethernet LAN. Some of these methods can be equally applied to token ring, but as Ethernet is currently the most popular technology it has been chosen in the analysis.

4.2 NETWORK TRAFFIC

Ethernet relies on nodes contending for access to the network and there is no priority given to any of the contending nodes. A sending node broadcasts frames in both directions on the network. It is the responsibility of the receiving node to accept, acknowledge, and reply to those transmissions. When more than one node tries to transmit at the same time a collision occurs and the contending nodes back-off for a random time period.

Under light traffic Ethernet is very efficient and has very fast processing, but is wasteful in the networks capacity. Under heavy traffic its efficiency decreases because there are likely to be more collisions. Research has shown that peak capacity of an Ethernet network occurs between 50% and 55% of rated capacity. Figure 4.1 shows an example of how the efficiency of the network varies with the amount of traffic. At low loadings the efficiency of the network is also low because much of the channels capacity is unused.

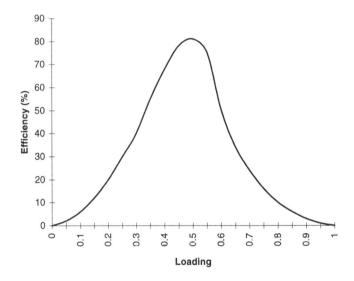

Figure 4.1   Network efficiency as a function of loading

4.3 IMPROVING NETWORK PERFORMANCE

The main methods that improve network performance are:

- to analyzing network traffic over time and implement changes to improve network loading;
- to verify that the network conforms to Ethernet specifications and minimize the number of different Ethernet standards;
- to isolate segments with network bridges;
- to remove network repeaters;
- to provide fan-out units;
- to use bridges which store frames and forward them when network traffic reduces.

4.3.1 Minimizing differing Ethernet standards

Most currently available Ethernet adapters conform to the IEEE 802.5 standard, but a degradation occurs when using different standards. Ethernet 1.0 does not work well with Ethernet 2.0 because of collision detection differences. Different implementations of software and/or architecture may also tend to reduce network performance.

4.3.2 Fan-out boxes

A fan-out box filters each of its clients and resolves potential collisions before their collisions reach the network backbone. It also provides a mini-network so that users on the same fan-out unit can communicate with each other without frames ever reaching the Ethernet backbone. Figure 4.2 illustrates two five-user fan-out units connected to a backbone. Heavy users are normally best concentrated together on the same fan-out. A fan-out box is also called a concentrator.

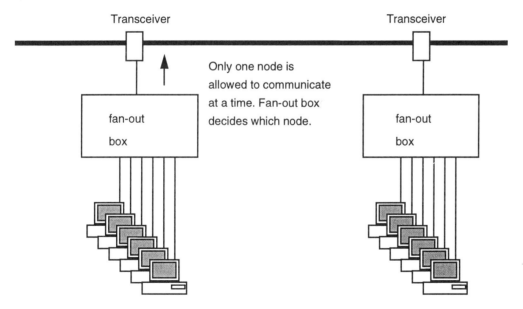

Figure 4.2 Fan-out units reduce network collisions

The fan-out construction leads to a network topology known as concentrated local area networks (CLANs). If correctly planned, such concentration substantially lowers network traffic, as Figure 4.3 shows, by facilitating precollision mediation.

Typically, fan-out boxes have 2, 4, 8 or 16 connections which can be cascaded to create a total fan-out of 64. They connect to the backbone using either by a standard Ethernet connection (that is, an N-type connection) or by a Cheapernet connection (that is, a BNC connection).

Each group, known as workgroups, can act independently from the rest of the network, if required. This is a major advantage as when the main backbone Ethernet cable fails then the connected units on the fan-out boxes can still communicate with each other. It can also be used where no backbone exists.

Figure 4.4 shows an example network with seven workgroups (MECH_1, INSTR_1, and so on). Each of the groups can have their own facilities such as a local file server, printer, and so on.

4.3.3 Segment network with bridges and fan-out boxes

Network bridges only allow traffic to flow out of a segment if the destination node is outside the network. Figure 4.5 shows an example of this. Fan-out boxes also segment networks.

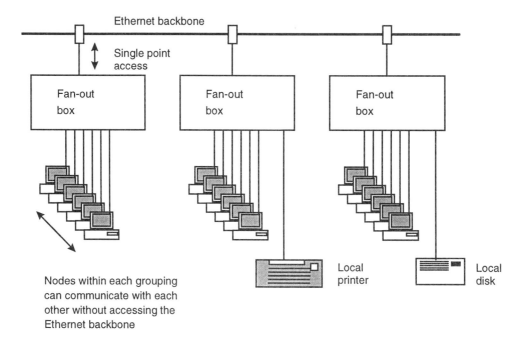

Figure 4.3   Fan-out units reduce network collisions

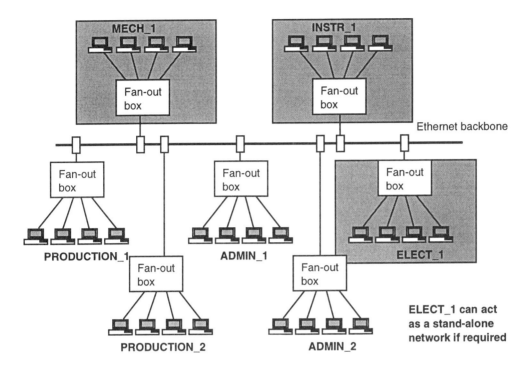

Figure 4.4   Organizing the network into workgroups

4.3.4 Distributing load

The network loading tends to vary over time, but it may vary consistently over a period of a day, week, month, year and so on. A sample daily loading is shown in Figure 4.5. This example shows that, for this network, the peak network traffic occurs at around 11 a.m. and also between 3 p.m. and 5 p.m. The network manager has tried to even-out network traffic by performing network back-ups at times when the network loading is low.

Much of the network traffic is due to disk transfers. By analyzing network statistics it is possible to determine when hot spots occur. From these statistics the network manager may ask some users to change the way they operate. For example, by staggering lunch breaks the network loading traffic could be evened-out. The system manager may also even-out the network traffic by only allowing certain applications (or users) to use the network at certain times of the day. It is typical for heavy processing or network intensive tasks to run in periods when there is a light network loading, typically at night.

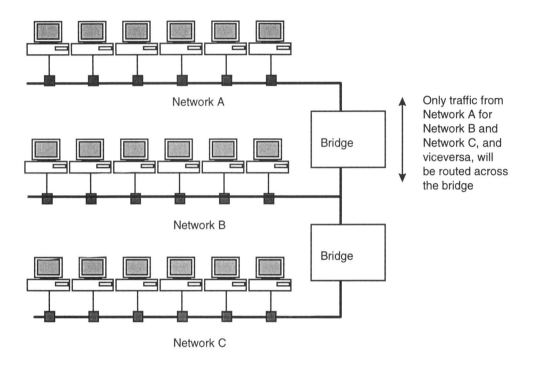

Only traffic from
Network A for
Network B and
Network C, and
viceversa, will
be routed across
the bridge

Figure 4.5   Segmentation using network bridges

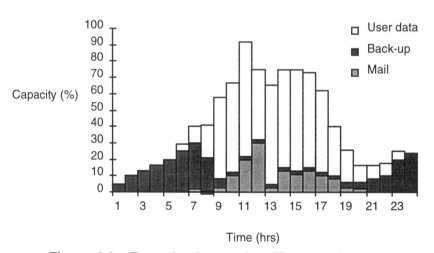

Figure 4.6   Example of network traffic over 24 hours

4.3.5 Provide alternative routes

Alternative network paths ease traffic on a heavily used channels. These
extra links could be direct Ethernet links (that is, not connected to the

main backbone) or could be through high speed optical or copper links. By identifying bottle-necks the number of collisions on the network can be significantly reduced.

Figure 4.7 shows an example of how adding an alternative route eases the traffic on a particular segment. Typically, peripherals which require large data transfer, such as image scanners, large storage disks, and so on, are isolated from the network and connected to the local bus of a computer or connected via high speed serial links.

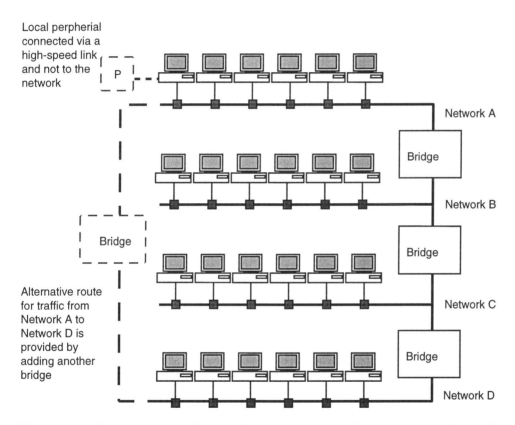

Figure 4.7 Example of adding a bridge to reduce traffic on network B and C

4.3.6 Remove repeaters

A repeater increases the length of a segment and thus increases the number of nodes which contend for the ether. This increases the chances of a collision which in turn reduces network traffic. Figure 4.8 shows an example of three segments connected by repeaters. A collision is more likely to occur as there are more nodes connected to the main Ethernet link. A 'jam' signal thus propagates over Network A, Network B and Network C. It is normally better to replace repeaters with network bridges.

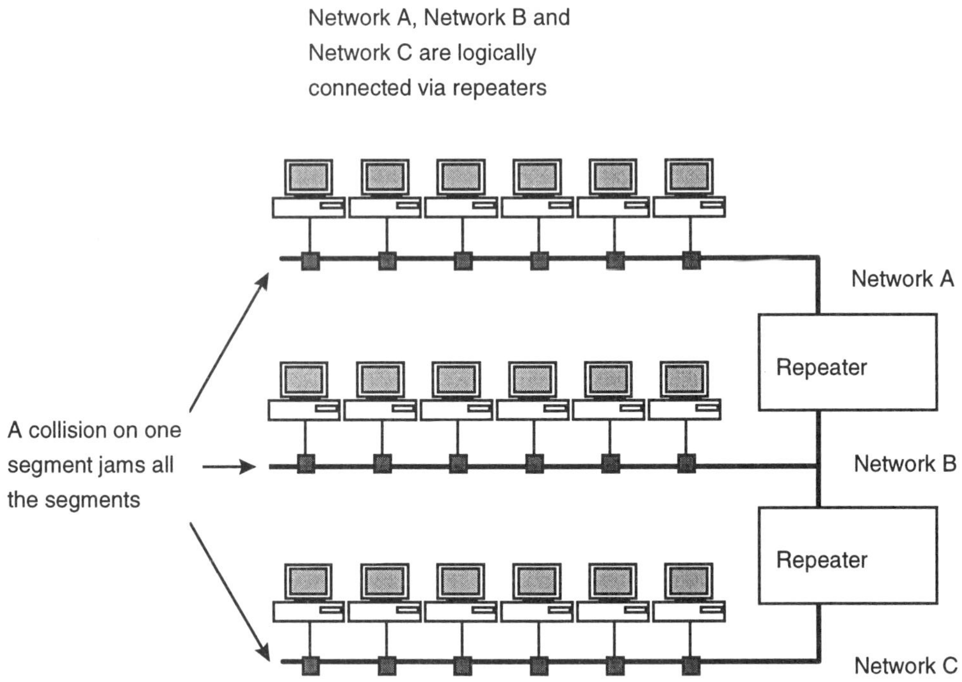

Figure 4.8 Repeaters connecting Network A, B and C

4.3.7 Provide local resources

Much of the traffic on the network is disk transfers, thus the traffic can be significantly reduced by providing local disk drives, which are local either to the nodes, segments or workgroups. Other resources such as printers can be also organised in this way to reduce the main backbone traffic. Figure 4.3 shows an example of local printers and disk drives.

4.3.8 Buffer networks with store and forward bridges

A buffer and store bridge store frames for a time and when the network is free it transmits them. This filtration process prevents collision and helps to even-out network traffic.

4.4 TUTORIAL

4.1 Explain the reason why the performance of an Ethernet network reduces as the traffic load becomes heavy.

4.2 Figure 4.9 shows the layout of five offices. There are 80 nodes in each of the offices connected to a single thin-wire Ethernet backbone. Discuss any possible problems with this network, such as network traffic, faults on the backbone, the number of nodes on the backbone, length of backbone, and so on.

4.3 Redesign the network in Figure 4.9 taking into account the problems raised in Q4.2.

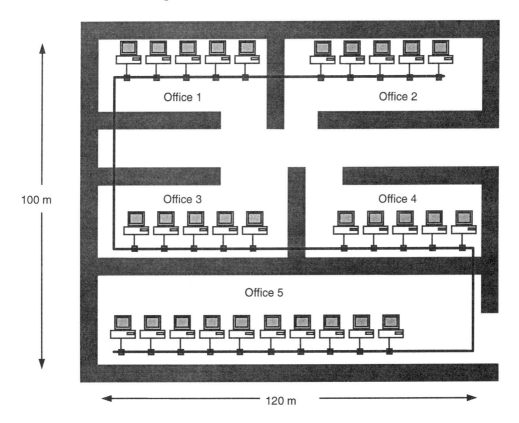

Figure 4.9   Office network plan

4.4 For a known network, such as a university or college network, determine any possible network problems and redesign it so that it works more efficiently. Also, if possible, estimate the number of nodes on each segment.

4.5   For a known network, such as a university or college network, make a simple estimation of the traffic loading on the network over a period of a day. The traffic estimate at the peak loading should be taken as 1 and all other loadings should be relative to this. Base the analysis on the number of users at a certain time of day and, if possible, the typical applications running at that time.

5

Transmission Control Protocol (TCP) and Internet Protocol (IP)

5.1 INTRODUCTION

The interconnection of networks is known as internetworking (or internet). Each part of an internet is a subnetwork (or subnet). TCP/IP are a pair of protocols which allow one subnet to communicate with another. A protocol is a set of rules which allow the orderly exchange of information. The IP part corresponds to the Network layer of the OSI model and the TCP part to the Transport layer. Their operation is transparent to the Physical and Data Link layers and can thus be used on Ethernet, FDDI or token ring networks.

TCP/IP was originally developed by the US Defence Advanced Research Projects Agency (DARPA). Their objective was to connect a number of universities and other research establishments to DARPA. The resultant internet is now known as the Internet, this is illustrated in Figure 5.1. It has since outgrown this application and many commercial organizations now connect to the Internet. The Internet uses TCP/IP to transfer data. Each node on the Internet is assigned a unique network address, called an IP address. Note that any organization can have its own internets, but if it is to connect to the Internet then the addresses must conform to the Internet addressing format.

The ISO have adopted TCP/IP as the basis for the standards relating to the network and transport layers of the OSI model. This standard is known as ISO-IP. Most currently available systems conform to the IP addressing standard.

Common applications that use TCP/IP communications are remote login and file transfer. Typical programs which can be run with this protocol are:

ftp   - file transfer program;

`telnet` - allows a user to log into another computer;

`ping` - a program to determine if a node is responding to TCP/IP communications.

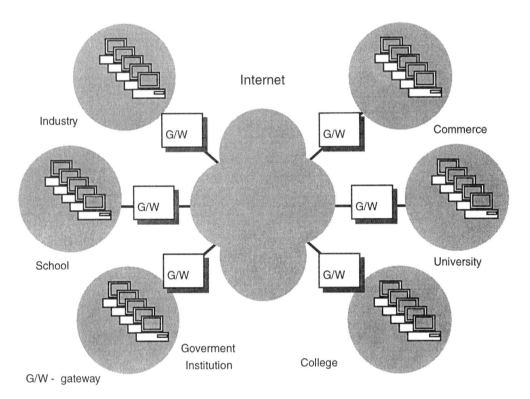

Figure 5.1   Internet via a gateway

A few advantages of internets are:

- they allow two or more networks to be managed as a single network, even when they are located over a large distance or within the same building;
- they allow one type of computer to communicate with another, possibly over large physical distances;
- they allow electronic mail to traverse through the internet;
- they allow multiple routes between nodes. This helps create alternative communication routes when links either are not operating or if they are busy;
- they have the capacity to isolate traffic from other networks;
- they allow access to information on remote sites;

- they improve efficiency and flexibility, such as providing work groups, isolating heavy traffic users, prevent access of certain users, and so on.

5.2 TCP/IP GATEWAYS AND HOSTS

TCP/IP hosts are nodes which communicate over interconnected networks using TCP/IP communications. A TCP/IP gateway node connects one type of network to another. It contains hardware to provide the physical link between the different networks and the hardware and software to convert frames from one network to the other. Typically, it converts a token ring MAC layer to an equivalent Ethernet MAC layer, and viceversa.

A router connects a network of a similar type to another of the same kind through a point-to-point link. The main operational difference between a gateway, a router, and a bridge, is that, for a token ring and Ethernet network, the bridge uses the 48-bit MAC address to route frames, whereas the gateway and router uses an IP network address. As an analogy to the public telephone system, the MAC address would be equivalent to a randomly assigned telephone number, whereas the IP address would contain the information on logically where the telephone is located, such as which country, area code, etc. A gateway and a router are illustrated in Figure 5.2.

Figure 5.3 shows how a gateway routes information. The gateway reads the frame from the computer on network A. It then reads the IP address contained in the frame and makes a decision whether it is routed out of network A to network B. If it does then it relays the frame to network B.

5.3 FUNCTION OF THE IP PROTOCOL

The main functions of the IP protocol are to:

- route IP data frames – which are called internet datagrams – around an internet. The IP protocol program running on each node knows the location of the gateway on the network. The gateway must then be able to locate the interconnected network. Data then passes from node to gateway through the internet;
- fragment the data into smaller units if it is greater than a given amount (64 kB);

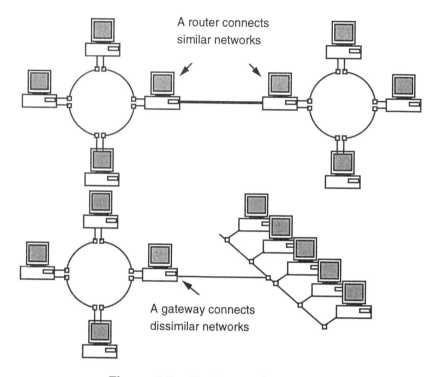

Figure 5.2   Routers and gateways

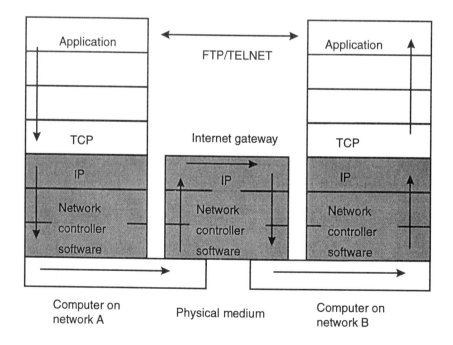

Figure 5.3   Internet gateway layers

- report errors. When a datagram is being routed or is being reassembled an error can occur. If this happen then the node that detects the error reports back to the source node. Datagrams are deleted from the network if they travel through the network for more than a set time. Again, an error message is returned to the source node to inform it that the internet routing could not find a route for the datagram or that the destination node, or network, does not exist.

5.4 INTERNET DATAGRAM

The IP protocol is an implementation of the network layer of the OSI model. It adds a data header onto the information passed from the transport layer, the resultant data packet is known as an internet datagram. The header contains information such as the destination and source IP addresses, the version number of the IP protocol and so on. Its format is given in Figure 5.4.

The datagram contains up to 65 536 bytes (64 kB) of data. If the data to be transmitted is less than, or equal to 64 kB, then it is sent as one datagram. If it is more than this then the source splits the data into fragments and sends multiple datagrams. When transmitted from the source each datagram is routed separately through the internet and the received fragments are finally reassembled at the destination.

The TCP/IP version number helps gateways and nodes interpret the data unit correctly. Differing version may have a different format or the IP protocol interprets the header differently.

The type of service bit field is an 8-bit bit pattern in the form PPPDTRXX. PPP defines the priority of the datagram (from 0 to 7), D sets a low delay service, T sets high throughput, R sets high reliability and XX are currently not used.

The header length defines the size of the data unit in multiplies of 4 bytes (32 bits). The minimum length is 5 bytes and the maximum is 65 536 bytes. Padding bytes fill any unused spaces.

A gateway may route a datagram and split it into smaller fragments. The D bit informs the gateway that it should not to fragment the data and thus signifies that a receiving node should receive the data as a single unit or not at all.

The M bit is the more fragments bit and is used when data is split into fragments. The fragment offset contains the fragment number.

A datagram could be delayed in the internet indefinitely. To prevent this the 8-bit time-to-live value is set to the maximum transit time in seconds. It is set initially by the source IP. Each gateway then decrements this value by a defined amount. When it becomes zero the datagram is discarded. It also defines the maximum amount of time that a destination IP node should wait for the next datagram fragment.

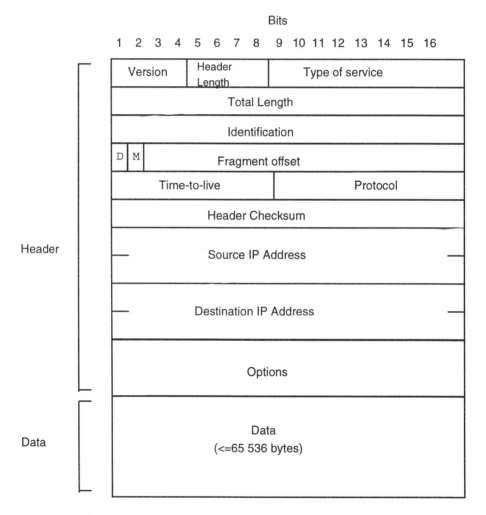

Figure 5.4   Internet datagram format and contents

Different IP protocols can be used on the datagram. The 8-bit proto-col field defines which type is to be used.

The header checksum contains a 16-bit pattern for error detection.

The source and destination IP addresses are stored in the

32-bit source and destination IP address fields. The `options` field contains information such as debugging, error control and routing information.

5.5 TCP/IP INTERNETS

Figure 5.5 illustrates a sample TCP/IP implementation. A gateway `MERCURY` provides a link between a token ring network (NETWORK A) and the Ethernet network (ETHER C). Another gateway `PLUTO` connects NETWORK B to ETHER C. The TCP/IP protocol allows a host on NETWORK A to communicate with `VAX01`.

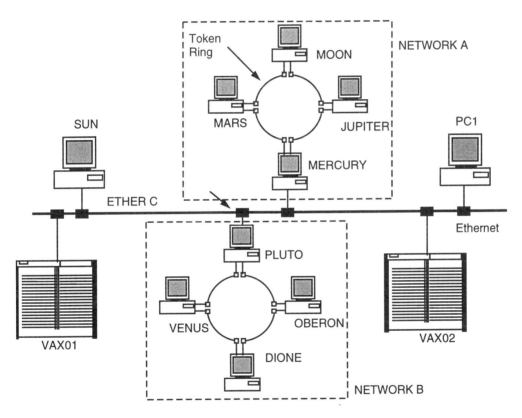

Figure 5.5   Example internet

5.5.1 Selecting internet addresses

Each node using TCP/IP communications requires a IP address which is then matched to its token ring or Ethernet MAC address. The MAC address allows nodes on the same segment to communicate with each other. In order for nodes on a different network to communicate, each must be configured with an IP address.

Nodes on a TCP/IP network are either hosts or gateways. Any nodes that run application software or are terminals are hosts. Any node which routes TCP/IP packets between networks is called a TCP/IP gateway node. This node must have the necessary network controller boards to physically interface to other networks it connects with.

5.5.2 Format of the IP address

A typical IP address consists of two fields: the left field (or the network numbcr) identifies the network, and the right number (or the host number) identifies the particular host within that network. Figure 5.6 illustrates this.

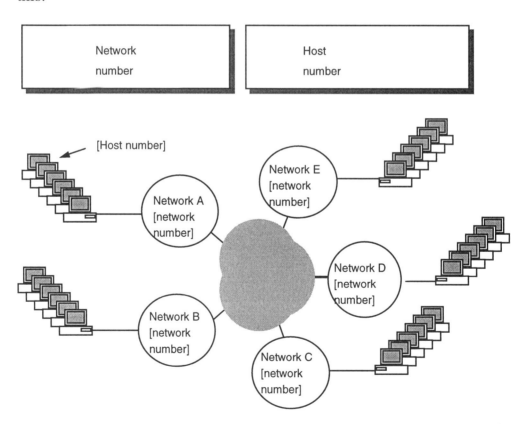

Figure 5.6   IP addressing over networks

The IP address is 32 bits long and can address over four million physical networks (2^{32} or 4 294 967 296 hosts). There are three different address formats, these are shown in Figure 5.7.

Each of these types is applicable to certain types of networks. Class A allows up to 128 (2^7) different networks and up to 16 777 216 (2^{24}) hosts on each network. Class B allows up to 16 384 networks and up to 65 536 hosts on each network. Class C allows up to 2 097 152 networks each with up to 256 hosts.

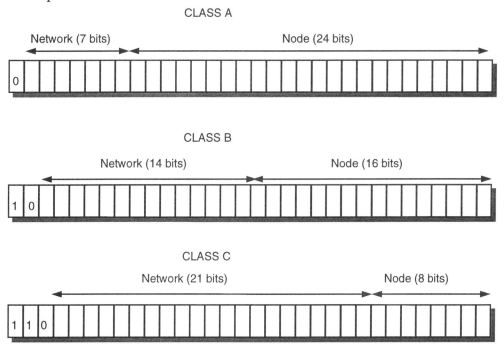

Figure 5.7   Type A, B and C IP address classes

The class A address is thus useful where there are a small number of networks with a large number of hosts connected to them. Class C is useful where there are many networks with a relatively small number of hosts connected to each network. Class B addressing gives a good compromize of networks and connected hosts.

When selecting internet addresses for the network, the address can be specified simply with decimal numbers within a specific range. The standard DARPA IP addressing format is of the form:

```
W.X.Y.Z
```

where W, X, Y and Z represent 1 byte of the IP address. As decimal numbers they range from 0 to 255. The 4 bytes together represent both the network and host address.

The valid range of the different IP addresses is given in Table 5.1. Thus for a class A type address there can be 127 networks and 16 711 680 (256×256×255) hosts. Class B can have 16 320 (64×255) networks and class C can have 2 088 960 (32×256×255) networks and 255 hosts.

Addresses above 223.255.254 are reserved, as are addresses with groups of zeros.

Table 5.1   Ranges of addresses for type A, B and C internet address

Type	Network portion	Host portion
A	1 - 126	0.0.1 - 255.255.254
B	128.1 - 191.254	0.1 - 255.254
C	192.0.1 - 223.255.254	1 - 254

5.5.3 Creating IP addresses with subnet numbers

Besides selecting IP addresses of internets and host numbers, it is also possible to designate an intermediate number called a subnet number. Subnets extend the network field of the IP address beyond the limit defined by the type A, B, C scheme. They allow a hierarchy of internets within a network. For example, it is possible to have one network number for a network attached to the internet, and various subnet numbers for each subnet within the network. This is illustrated in Figure 5.8.

For an address X.Y.Z.W and type for a type A address X specifies the network and Y the subnet. For type B the Z field specifies the subnet.

To connect to a global network a number is normally assigned by a central authority. For the Internet network it is assigned by the Network Information Centre (NIC). Typically, on the Internet an organization is assigned a type B network address. The first two fields of the address specify the organization network, the third specifies the subnet within the organization and the final specifies the host.

5.5.4 Specifying subnet masks

If a subnet is used then a bit mask, or subnet mask, must be specified to show which part of the address is the network part and which is the host.

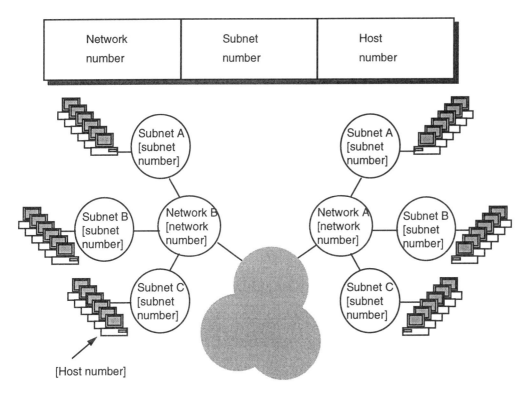

Figure 5.8 IP addresses with subnets

The subnet mask is a 32-bit number which has 1's for bit positions specifying the network and subnet parts and 0's for the host part. A text file called *hosts* is normally used to set up the subnet mask. Table 5.2 shows example subnet masks.

Table 5.2 Default subnet mask for type A, B and C IP addresses

Address Type	Default mask
Class A	255.0.0.0
Class B	255.255.0.0
Class C and Class B with a subnet	255.255.255.0

To set up the default mask the following line is added to the *hosts* file.

```
Hosts file
255.255.255.0 defaultmask
```

The following are the steps required to set up TCP/IP networks:

1. Host names are assigned easily remember names, such as names of towns, planets, moons, star systems, etc. It is easier to recall the name of a host as a planet name than to recall its IP address, e.g. `mars` is easier to remember than `146.176.147.43`.

2. The type of IP addressing is assigned. If there are a large number of hosts on a few networks then type A or B addressing is used. If there are many networks and with relatively few hosts then type C addressing is used.

    To connect to a global network, such as the Internet, a network address is granted, normally this is a type B address with a subnets field.

3. A network number is assigned. If the network number is not already assigned then the IP address is selected which is unique to all connected networks. To connect to a global network, such as the Internet, an address is assigned by a central authority, such as the NIC. An example of this is shown in Figure 5.9.

4. Subnet numbers are assigned.

Network number (X.Y)	Subnet number (Z)	Host number (W)

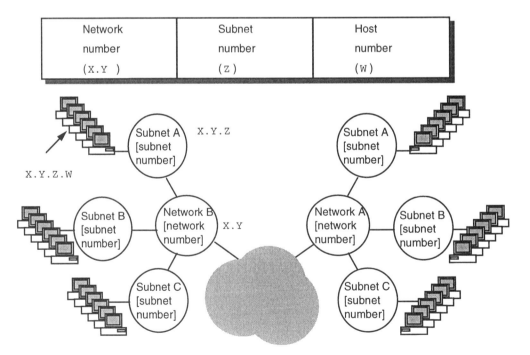

Figure 5.9   Internet addresses with subnets

5. Internet addresses for each host are assigned. Each host on a subnet must have the same network address. A file named *hosts* contains the aliases for the nodes on the network and their related IP addresses.

    On Unix systems this file can be found in the /etc directory. A sample hosts file is shown next. In this case dione, saturn, mercury and earth are host numbers 30, 31, 32 and 33 respectively on the subnet 146.176.147.

```
▤ Hosts file
146.176.147.30   dione
146.176.147.31   saturn
146.176.147.32   mercury
146.176.147.33   earth
```

6. Gateways are assigned internet addresses. A gateway connects to two different networks thus it must have two different IP addresses. In the *hosts* files given next it can be seen that the gateway oberon connects the subnets 146.176.146 and 146.176.145.

```
▤ Extract from a Hosts file
146.176.144.20   triton
146.176.146.23   oberon
146.176.145.23   oberon
```

5.6 EXAMPLE NETWORK

A university network is shown in Figure 5.10. The connection to the outside global Internet network is via the Janet gateway node, its IP address is 146.176.1.3. Three subnets 146.176.160, 146.176.129 and 146.176.151 connect the gateway to departmental bridges. The Computer Studies bridge address is 146.176.160.1 and the Electrical department bridge has an address 146.176.151.254.

The Electrical department bridge links, through other bridges, to the subnets   146.176.144,   146.176.145,   146.176.147, 146.176.150 and 146.176.151.

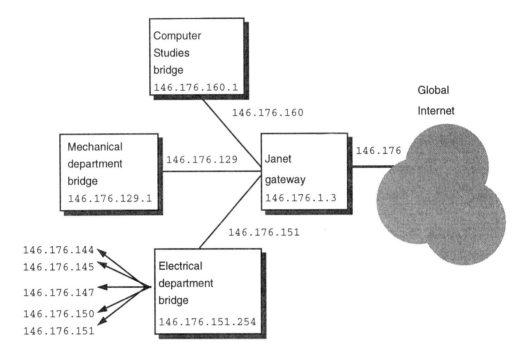

Figure 5.10 Example network

The topology of the Electrical department network is shown in Figure 5.11. The main bridge into the department connects to two Ethernet network of PCs (subnets 146.176.150 and 146.176.151) and to another bridge (Bridge 1). Bridge 1 connects to the subnet 146.176.144. Subnet 146.176.144 connects to workstations and X-terminals. It also connects to the gateway moon which links the token ring subnet 146.176.145 with the Ethernet subnet 146.176.144. The gateway Oberone, on the 146.176.145 subnet, connects to an Ethernet link 146.176.146. This then connects to the gateway Dione which is also connected to the token ring subnet 146.176.147.

Each node on the network is assigned with an IP address. The *hosts* file for the set up in Figure 5.11 is shown next. For example the IP address of mimas is 146.176.145.21 and for miranda it is 146.176.144.14. It should be noticed that the gateway nodes: oberon, moon and dione all have two IP addresses.

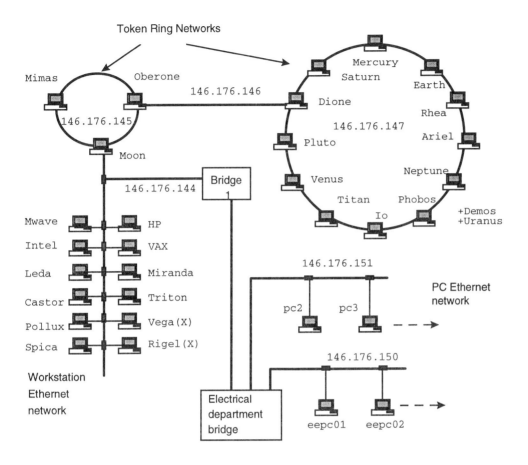

Figure 5.11 Example network

📄 Contents of host file

```
146.176.1.3         janet
146.176.144.10      hp
146.176.145.21      mimas
146.176.144.11      mwave
146.176.144.13      vax
146.176.144.14      miranda
146.176.144.20      triton
146.176.146.23      oberon
146.176.145.23      oberon
146.176.145.24      moon
146.176.144.24      moon
146.176.147.25      uranus
146.176.146.30      dione
146.176.147.30      dione
146.176.147.31      saturn
146.176.147.32      mercury
146.176.147.33      earth
146.176.147.34      deimos
146.176.147.35      ariel
```

```
146.176.147.36        neptune
146.176.147.37        phobos
146.176.147.39        io
146.176.147.40        titan
146.176.147.41        venus
146.176.147.42        pluto
146.176.147.43        mars
146.176.147.44        rhea
146.176.147.22        jupiter
146.176.144.54        leda
146.176.144.55        castor
146.176.144.56        pollux
146.176.144.57        rigel
146.176.144.58        spica
146.176.151.254       cubridge
146.176.151.99        bridge_1
146.176.151.98        pc2
146.176.151.97        pc3
            :::::
146.176.151.71        pc29
146.176.151.70        pc30
146.176.151.99        ees99
146.176.150.61        eepc01
146.176.150.62        eepc02
255.255.255.0         defaultmask
```

5.7 DOMAIN NAME SYSTEM

An IP address can be defined in the form XXX.YYY.ZZZ.WWW, where XXX, YYY, ZZZ and WWW are integer value in the range 0 to 255. On the Internet the XXX.YYY.ZZZ part normally defines the subnet and the WWW the host. Such names may be difficult to remember. A better method is to use symbolic names rather than IP addresses.

Users and application programs can then use symbolic names rather than IP addresses. The directory network services on the Internet determines the IP address of the named destination user or application program. This has the advantage that users and application program can move around the Internet and are not fixed to an IP address.

An analogy relates to the public telephone service. A phone directory contains a list of subscribers and their associated telephone number. If someone looks for a telephone number, first the user's name is looked-up and their associated phone number found. The telephone directory listing maps a users name (symbolic name) to an actual telephone number (the actual address).

Table 5.3 lists some example Internet domain assignments for World Wide Web (WWW) servers. Note that domain assignments are not fixed and can change their corresponding IP addresses, if required. The binding between the symbolic name and its address can thus change at any time.

Table 5.3 Example Internet network addresses

Web server	*Internet domain names*	*Internet IP address*
NEC	web.nec.com	143.101.112.6
Sony	www.sony.com	198.83.178.11
Intel	www.intel.com	134.134.214.1
IEEE	www.ieee.com	140.98.1.1
University of Bath	www.bath.ac.uk	136.38.32.1
University of Edinburgh	www.ed.ac.uk	129.218.128.43
IEE	www.iee.org.uk	193.130.181.10
University of Manchester	www.man.ac.uk	130.88.203.16

5.8 INTERNET NAMING STRUCTURE

The Internet naming structure uses labels separated by periods, an example is eece.napier.ac.uk. It uses a hierarchical structure where organizations are grouped into primary domain names. These are com (for commercial organizations), edu (for educational organizations), gov (for government organizations), mil (for military organizations), net (Internet network support centres) or org (other organizations). The primary domain name may also define which country the host is located, such as usa (USA), uk (United Kingdom), fr (France), and so on. All hosts on the Internet must be registered to one of these primary domain names.

The labels after the primary field relate the subnets within the network. For example in the address eece.napier.ac.uk, the ac label relates to an academic institution within the uk, napier to the name of the institution and eece the subnet with that organization.

In the example in Figure 5.12 there are four labels below the ac.uk group, these are ed (University of Edinburgh), bath (University of Bath), napier (Napier University) and man (University of Manchester). There is a subnet on the napier address called eece. On this network there is a host computer called www (the WWW server). The full Internet name of this computer is thus www.eece.napier.ac.uk, the directory services will map this to an Internet IP address.

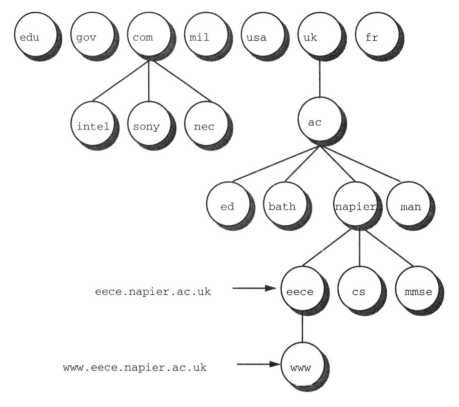

Figure 5.12   Example domain naming

5.9 DOMAIN NAME SERVER

Each institution on the Internet has a host which runs a process called the domain name server (DNS). The DNS maintains a database called the directory information base (DIB) which contains directory information for that institution. When a new host is added, the system manager adds its name and its IP address. It can then access the Internet.

5.10 TCP/IP COMMANDS

There are several standard programs available over TCP/IP links. The example session is this section relate to the network outlined in Figure 5.11.

5.10.1 ping

The `ping` program determines if a node is responding to TCP/IP communication. Sample run 5.1 shows that miranda is active and ariel isn't.

🖥 Sample run 5.1: Using PING command
```
C:\WINDOWS>ping miranda
miranda (146.176.144.14) is alive
C:\WINDOWS>ping ariel
no reply from ariel (146.176.147.35)
```

5.10.2 ftp (file transfer protocol)

The ftp program uses the TCP/IP protocol to transfer files to and from remote nodes. If necessary, it reads the *hosts* file to determine the IP address. Once the user has logged into the remote node the commands that can be used are similar to DOS commands such as cd (change directory), dir (list directory), open (open node), close (close node), pwd (present working directory). The get command copies a file from the remote node and the put command to copy it to the remote node.

The type of file to be transferred must also be specified. This file can be ASCII text (the command ascii) or binary (the command binary).

Sample run 5.2 shows a session with the remote VAX computer (Internet name VAX, address 146.176.144.13). The get command is used to get the file *TEMP.DOC* from VAX and transfer it to the calling PC.

🖥 Sample run 5.2: Using FTP to get files from a remote site (e.g. VAX)
```
C:\NET> ftp vax
Connected to vax.
Name (vax:nobody): bill_b
Password (vax:bill_b):
331 Password required for bill_b.
230 User logged in, default directory DUA2:[STAFF.BILL_B]
ftp> dir
200 PORT Command OK.
125 File transfer started correctly
commands.dir;1    MAY 10 11:00 1990       512 (,RWE,RWE,RE)
docs.dir;1        MAY  4 13:31 1993       512 (,RWE,RWE,RE)
fortran.dir;1     MAY 10 11:00 1990       512 (,RWE,RWE,RE)
login.com;29      MAY 10 12:14 1994      1044 (,RWE,RE,RE)
mail.mai;1        MAY  4 10:58 1993     15360 (,RW,,RE)
pascal.dir;1      MAY 10 11:00 1990       512 (,RWE,RWE,RE)
temp.doc;1        MAY  5 07:33 1993        46 (,RWE,RE,)
226 File transfer completed ok
754 bytes received in 2.012100 seconds (0.37 Kbytes/s)
ftp>
```

```
ftp> get temp.doc
200 PORT Command OK.
125 File transfer started correctly
226 File transfer completed ok
45 bytes received in 0.005000 seconds (8.79 Kbytes/s)
ftp> quit
221 Goodbye.

C:\NET>dir *.doc

 Volume in drive C is MS-DOS_5
 Volume Serial Number is 3B33-13D3
 Directory of C:\NET

ASKME     DOC      3369 03/07/92    1:25
TEMP      DOC        45 24/05/94   14:47
        2 file(s)       3414 bytes
                     2093056 bytes free
C:\NET>
```

Sample run 5.3 shows a session of sending a file from the local node (in this case the PC) and to a remote node (in this case VAX). The put command is used for this purpose.

⌨ Sample run 5.3: Using FTP to send files to a remote site
```
C:\NET>ftp vax
Connected to vax.
Name (vax:nobody): bill_b
Password (vax:bill_b):
331 Password required for bill_b.
230 User logged in, default directory DUA2:[STAFF.BILL_B]
ftp> put askme.doc
200 PORT Command OK.
125 File transfer started correctly
226 File transfer completed ok
3369 bytes sent in 0.011000 seconds (299.09 Kbytes/s)
ftp> dir *.doc
200 PORT Command OK.
125 File transfer started correctly
askme.doc;1    MAY 24 14:53 1994      3396 (,RW,R,R)
temp.doc;1     MAY  5 07:33 1993        46 (,RWE,RE,)
226 File transfer completed ok
215 bytes received in 1.019100 seconds (0.21 Kbytes/s)
ftp>
```

5.10.3 telnet

The telnet program uses TCP/IP to remotely log into a remote node. Sample run 5.4 shows an example of login into the node miranda.

⌨ Sample run 5.4: Using TELNET for remote login
```
C:\NFS>telnet miranda
      HP-UX miranda A.09.01 A 9000/720 (ttys5)
login: bill_b
Password:
      (c)Copyright 1983-1992 Hewlett-Packard Co.,  All Rights Reserved.
            :::::::
      (c)Copyright 1988 Carnegie Mellon
[51:miranda :/net/castor_win/local_user/bill_b ] %
```

5.10.4 nslookup

The nslookup program interrogates the local hosts file and the DNS to determine the IP address of an Internet node. If it cannot find it in the local file then it communicates with gateways outside its own network to see if they know the address. Sample run 5.5 shows that the IP address of the WWW server is 134.134.214.1.

⌨ Sample run 5.5: Example of nslookup
```
C:\> nslookup
Default Server:   ees99.eece.napier.ac.uk
Address:   146.176.151.99
> www.intel.com
Server:   ees99.eece.napier.ac.uk
Address:   146.176.151.99
Name:     web.jf.intel.com
Address:   134.134.214.1
Aliases:  www.intel.com
```

5.10.5 netstat (network statistics)

On a Unix system the command netstat can be used to determine the status of the network. The -r option shown in sample run 5.6 shows that this node uses moon as a gateway to another network.

⌨ Sample run 5.6: Using Unix netstat command
```
[54:miranda :/net/castor_win/local_user/bill_b ] % netstat -r
Routing tables
Destination         Gateway            Flags    Refs     Use
Interface
localhost           localhost          UH        0      27306   lo0
default             moon               UG        0    1453856   lan0
146.176.144         miranda            U         8    6080432   lan0
146.176.1           146.176.144.252    UGD       0         51   lan0
146.176.151         146.176.144.252    UGD      11       5491   lan0
[55:miranda :/net/castor_win/local_user/bill_b ] %
```

5.10.6 route

Most bridges build up a picture of the interconnected networks. For this purpose the bridge sets up a routing table. On a Unix system the routing table can be modified manually using the `route` program. For PC-NFS a it is specified using the `net route` command. This is shown in sample run 5.7. All Internet addresses which are not local to the segment will be routed via `cubbridge`.

🖳 Sample run 5.7: Routing a gateway with PC-NFS
```
C:\WINDOWS>net route cubridge
C:\WINDOWS>net route
Non-local routing via gateway cubridge (146.176.151.254).
```

On a Unix system the routing can be specified by the `route` command. To add a gateway the `route add net cubridge gateway` could be used to specify that `cubridge` (`146.176.151.254`) is to be used as a gateway. Refer to a Unix manual for more help.

5.11 EXERCISE

5.1   Which of the following best describes a gateway:

		✓

A   it speeds the routing frames on a network
B   it connects a network to a telephone line
C   it echo's all frames from one network to another
D   it connects dissimilar networks

5.2   The `ping` program is used to:

		✓

A   determine if a node is responding to TCP/IP communications
B   determine IP addresses
C   remotely log into a node
D   transfer files from a remote node

5.3   The `telnet` program is used to:

A   determine if a node is responding to TCP/IP communications
B   remotely log into a node
C   transfer files from a remote node
D   determine IP addresses

5.4   The `ftp` program is used to:

A   determine if a node is responding to TCP/IP communications
B   remotely log into a node
C   transfer files from a remote node
D   determine IP addresses

5.12 TUTORIAL

5.5   Determine the IP addresses, and their type, of the following 32-bit addresses:

(i)   `10001100.01110001.00000001.00001001`
(ii)   `01000000.01111101.01000001.11101001`
(iii) `10101110.01110001.00011101.00111001`

5.6   If possible, determine some IP addresses and their corresponding Internet domain names.

5.7   Determine the countries which use the following primary domain names:

(a) `de`   (b) `nl`   (c) `it`   (d) `se`   (e) `dk`   (f) `sg`
(g) `ca`   (h) `ch`   (i) `tr`   (j) `jp`   (k) `au`

Determine some other domain names.

5.8   For a known TCP/IP network determine the names of the nodes and their Internet addresses.

6

High-level Data Link Control (HDLC)

6.1 INTRODUCTION

The data link layer is the second layer in the OSI seven layer model and its protocols define rules for the orderly exchange of data information between 2 adjacent nodes connected by a data link. Final framing, flow control between nodes, and error detection and correction are added at this layer. In previous chapters the data link layer was discussed in a practical manner. In this chapter its functions will be discussed with reference to HDLC.

The two types of protocol are:

- asynchronous protocol;
- synchronous protocol.

Asynchronous communications uses start-stop method of communication where characters are sent between nodes, as illustrated in Figure 6.1. Special characters are used to control the data flow. Typical flow control characters are End of Transmission (EOT), Acknowledgement (ACK), Start of Transmission (STX) and Negative Acknowledgement (NACK).

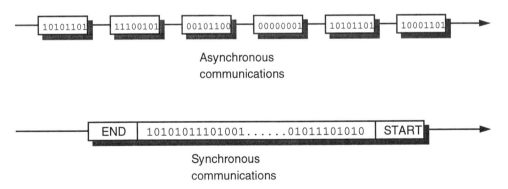

Figure 6.1   Asynchronous and synchronous communications

Synchronous communications involves the transmission of frames of bits with start and end bit characters to delimit the frame. The most popular are IBM's synchronous data link communication (SDLC) and high-level data link control (HDLC). Many network data link layers are based upon these standards, examples include the LLC layer in IEE 802.*x* LAN standards and LAPB in the X.25 packet switching standard.

Synchronous communications normally uses a bit-oriented protocol (BOP), where data is sent one bit at a time. The data link control information is interpreted on a bit-by-bit basis rather than with unique data link control characters.

HDLC is a standard developed by the ISO to provide a basis for the data link layer for point-to-point and multi-drop connections. It can transfer data either in a simplex, half-duplex, or full-duplex mode. Frames are generally limited to 256 bytes in length and a single control field performs most data link control functions.

6.2 HDLC PROTOCOL

In HDLC, a node is either defined as a primary station or a secondary station. A primary station controls the flow of information and issues commands to secondary stations. The secondary station then sends back responses to the primary. A primary station with one or more secondary stations is known as unbalanced configuration.

HDLC allows for point-to-point and multi-drop. In point-to-point communications a primary station communicates with a single secondary station. For multi-drop, one primary station communications with many secondary stations.

In point-to-point communications it is possible for a station be operate as a primary and a secondary station. At any time one of the stations can be a primary and the other the secondary. Thus commands and responses flow back and forth over the transmission link. This is known as a balanced configuration, or combined stations.

6.2.1 HDLC modes of operation

HDLC has three modes of operation. Unbalanced configurations can use the normal response mode (NRM). Secondary stations can only transmit when specifically instructed by the primary station. When used as a point-to-point or multi-drop configuration only one primary station is used. Figure 6.2 shows a multi-drop NRM configuration.

Unbalanced configurations can also use the asynchronous response mode (ARM). It differs from NRM in that the secondary is allowed to communicate with the primary without receiving permission from the primary.

In asynchronous balanced mode (ABM) all stations have the same priority and can perform the functions of a primary and secondary station.

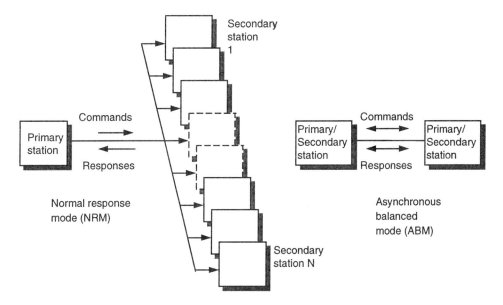

Figure 6.2 NRM and ABM mode

6.2.2 HDLC frame format

HDLC frames are delimited by the bit sequence `01111110`. Figure 6.3 shows the standard format of the HDLC frame, the 5 fields are the:

* flag field;
* address field;
* control field;
* information field;
* frame check sequence (FCS) field.

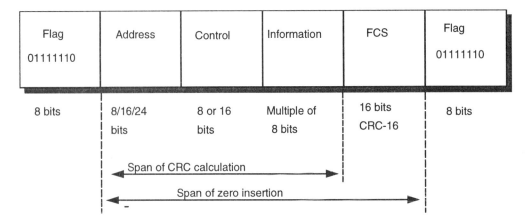

Figure 6.3   HDLC frame structure

6.2.3 Information field

The information fields contain data, such as OSI level 3, and above, information. It contains an integer number of bytes and thus the number of bits contained is always a multiple of eight. The receiver determines the number of bytes in the data because it can detect the start and end flag. By this method it also finds the FCS field. Note that the number of characters in the information can be zero as not all frames contain data.

6.2.4 Flag field

A unique flag sequence, 01111110 (or 7Eh), delimits the start and end of the frame. As this sequence could occur anywhere within the frame a technique called bit-insertion is used to stop this happening except at the start and end of the frame.

6.2.5 Address field

The address field is used to address connected stations an, in basic addressing, it contains an 8-bit address. It can also be extended, using extended addressing, to give any multiple of 8 bits.

When it is 8 bits wide it can address up to 254 different nodes, as illustrated in Figure 6.4. Two special addresses are 00000000 and 11111111. The 00000000 address defines the null or void address and the 11111111 broadcasts a message to all secondaries. The other 254 addresses are used to address secondary nodes individually.

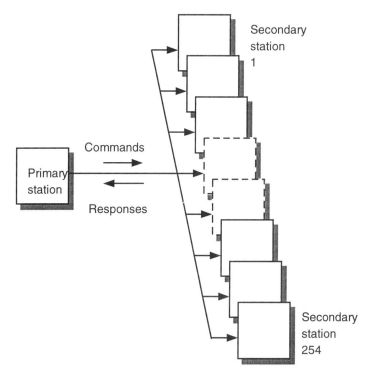

Figure 6.4   HDLC addressing range

If there are a large number of secondary stations then extended address can be used to extend the address field indefinitely. A 0 in the first bit of the address field allows a continuation of the address, or a 1 ends it. For example:

        XXXXXXX1 XXXXXXX0 XXXXXXX0 XXXXXXX0

6.2.6 Control field

The control field can either be 8 or 16 bits wide. It is used to identify the frame type and can also contain flow control information. The first two bits of the control field define the frame type, as shown in Figure 6.5. There are three types of frames, these are:

• information frames;

- supervisory frames;
- unnumbered frames.

When sent from the primary the P/F bit indicates that it is polling the secondary station. In an unbalanced mode a secondary station cannot transmit frames unless the primary sets the poll bit.

When sending frames from the secondary, the P/F bit indicates whether the frame is the last of the message, or not. Thus if the P/F bit is set by the primary it is a poll bit (P), if it is set by the secondary it is a final bit (F).

The following sections describe 8-bit control fields. Sixteen-bit control fields are similar but reserve a 7-bit field for the frame counter variables N(R) and N(S).

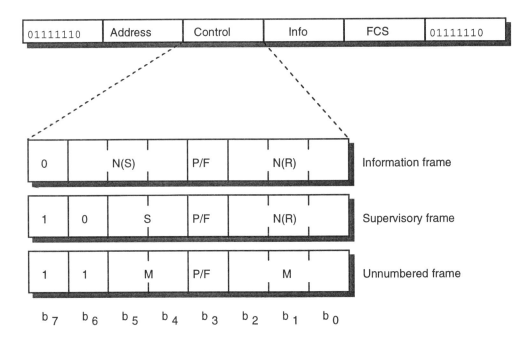

Figure 6.5 Format of a 8-bit control field

Information frame
An information frame contains sequenced data and is identified by a 0 in the first bit position of the control field. The 3-bit variable N(R) is used to confirm the number of transmitted frames received correctly and N(S) is used to number an information frame. The first frame transmitted is

numbered 0 as (000), the next as 1 (001), until the eighth which is numbered 111. The sequence then starts back at 0 again and this gives a sliding window of eight frames.

Supervisory frame

Supervisory frames contain flow control data. They confirm or reject previously received information frames and also can indicate whether a station is ready to receive frames.

The N(S) field is used with the S bits to acknowledge, or reject, previously transmitted frames. Responses from the receiver are set in the S field, these are receiver ready (RR), ready not to receive (RNR), reject (REJ) and selectively reject (SREJ). Table 6.1 gives the format of these bits.

Table 6.1 Supervisory bits

b_5	b_4	*Receiver status*
0	0	Receiver ready (RR)
1	0	Receiver not ready (RNR)
0	1	Reject (REJ)
1	1	Selectively reject (SREJ)

RR informs the receiver that it acknowledges the frames sent up to N(R). RNR tells the transmitter that the receiver cannot receive any more frames at the present time (RR will cancel this). It also acknowledges frames up to N(R). The REJ control rejects all frames after N(R). The transmitter must then send frames starting at N(R).

Unnumbered frame

If the first two bits of the control field are 1's then it is an unnumbered frame. Apart from the P/F flag the other bits are used to send unnumbered commands. When sending commands, the P/F flag is a poll bit (asking for a response), and for responses it is a flag bit (end of response).

The available commands are SARM (set asynchronous response mode), SNRM (set normal response mode), SABM (set asynchronous balance mode), RSET (reset), FRMR (frame reject) and Disconnect (DISC). The available responses are UA (unnumbered acknowledge), CMDR (command reject), FRMR (frame reject) and DM (disconnect mode). Bit definitions for some of these are:

```
SABM   1111P110   DM     1111F000   DISC   1100P010
UA     1100F110   FRMR   1110F001
```

6.2.7 Frame check sequence field

The frame check sequence (FCS) field contains an error detection code based on cyclic redundancy check (CRC) polynomials. It is used to check the address, control and information fields, as previously illustrated in Figure 6.2. HDLC uses a polynomial specified by CCITT V.41, which is $G(x) = x^{16} + x^{12} + x^5 + x^1$. This is also known as CRC-16 or CRC-CCITT. Refer to Chapter 13 for more information on CRCs.

6.3 TRANSPARENCY

The flag sequence 01111110 can occur anywhere in the frame. To prevent this a transparency mechanism called zero-bit insertion or zero stuffing is used. There are two main rules that are applied, these are:

- in the transmitter, a 0 is automatically inserted after five consecutive 1's, except when the flag occurs;
- at the receiver, when five consecutive 1's are received and the next bit is a 0 then the 0 is deleted and removed. If it is a 1 then it must be a valid flag.

In the following example a flag sequence appears in the data stream where it is not supposed to (spaces have been inserted around it). Notice that the transmitter detects five 0's in a row and inserts a 0 to break them up.

```
Message: 00111000101000 01111110   01011111 1111010101
Sent:    00111000101000 011111010 010111110111010101
```

6.4 FLOW CONTROL

Supervisory frames (S[]) send flow control information to acknowledge the reception of data frames or to reject frames. Unnumbered frames

(U[]) set up the link between a primary and a secondary, by the primary sending commands and the secondary replying with responses. Information frames (I[]) contain data.

6.4.1 Link connection

Figure 6.6 shows how a primary station (node A) sets up a connection with a secondary station (node B) in NRM (normal response mode). In this mode one or many secondary stations can exist. First the primary station requests a link by sending an unnumbered frame with: node B's address (ADDR_B), the set normal response mode (SNRM) command and with poll flag set (P=1), that is, U[SNRM,ABBR_B,P=1]. If the addressed secondary wishes to make a connection then it replies back with an unnumbered frame containing: its own address (ADDR_B), the unnumbered acknowledge (UA) response and the final bit set (F=1), i.e. U[UA,ABBR_B,F=1]. The secondary sends back its own address because many secondaries can exist and it thus identifies which station has responded. There is no need to send the primary station address as only one primary exists.

Once the link is set up data can flow between the nodes. To disconnect the link, the primary station sends an unnumbered frame with: node B's address (ADDR_B), the disconnect (DISC) command and the poll flag set (P=1), that is, U[DISC,ABBR_B,P=1]. If the addressed secondary accepts the disconnection then it replies back with an unnumbered frame containing: its own address (ADDR_B), the unnumbered acknowledge (UA) response and the final bit set (F=1), i.e. U[UA,ABBR_B,F=1].

When two stations act as both primaries and secondaries then they use the asynchronous balanced mode (ABM). Each station has the same priority and can perform the functions of a primary and secondary station. Figure 6.7 shows a typical connection. The ABM mode is set up initially using the SABM command (U[SABM,ABBR_B,P=1]). The connection between node A and node B is then similar to the NRM but, as node B operates as a primary station, it can send a disconnect command to node A (U[DISC,ABBR_B,P=1]).

The SABM, SARM and SNRM modes set up communications using an 8-bit control field. Three other commands exist which set up a 16-bit control field, these are SABME (set asynchronous balanced mode extended), SARME and SNRME. The format of the 16-bit control field is given in Figure 6.8.

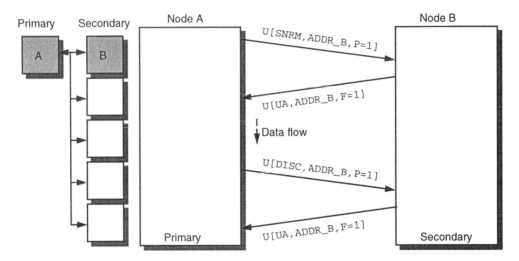

Figure 6.6   Connection between a primary and secondary in NRM

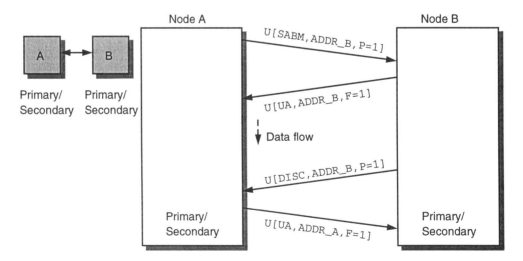

Figure 6.7   Connection between a primary/secondary in SABM

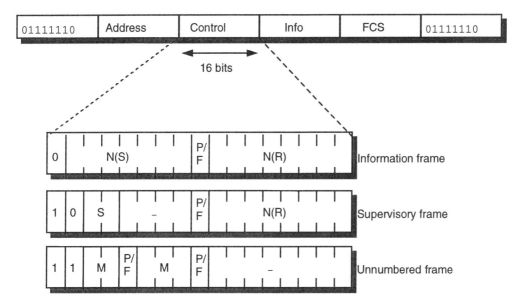

Figure 6.8 Extended control field

6.4.2 Data flow

After the link has been established then data can flow between two nodes. Information frames (`I[]`) send data and supervisory frames (`S[]`) provide flow control information to acknowledge the reception of data frames or to reject frames.

An information frame sends two sequence numbers in the control field `I[N(S),N(R)]`. For an 8-bit control field `N(S)` and `N(R)` are 3-bit counter values in the range of 0 to 7. The `N(S)` number is the sequence number of the information frame and `N(R)` is the sequence number for the frame the station expects to receive next. A supervisory frame has only the `N(R)` sequence number its field (`S[N(R)]`).

Figure 6.9 shows an example conversation between a sending station (node A) and a receiving station (node B). Initially three information frames are sent numbered 2, 3 and 4 (`I[N(S)=2]`, `I[N(S)=3]` and `I[N(S)=4, P=1]`). The last of these frames has the poll bit set, which indicates to node B that node A wishes it to respond, either to acknowledge or reject previously unacknowledged frames. Node B does this by sending back a supervisory frame (`S[RR, N(R)=5]`) with the receiver ready (RR) acknowledgement. This informs node A that node B expects to receive frame number 5 next. Thus it has acknowledged all frames up to and including frame 4.

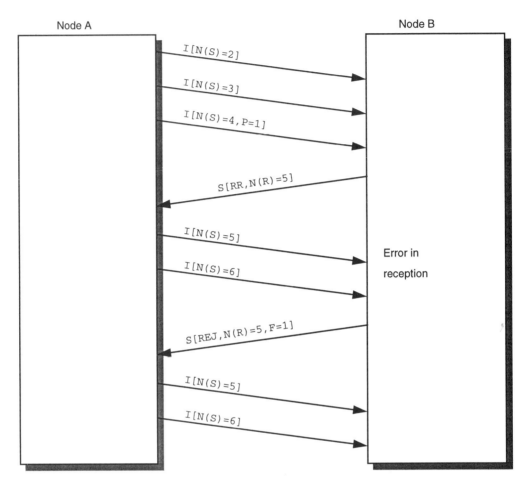

Figure 6.9 Example flow

In the example in Figure 6.9 an error has occurred in the reception of frame 5. The recipient informs the sender by sending a supervisory frame with a reject flow command (S[REJ, N(R)=5]). After the sender receives this it resends each frame after and including frame 5.

If the receiver does not want to communicate, at the present, it sends a receiver not ready flow command. For example S[RNR, N(R)=5] tells the transmitter to stop sending data, at the present. It also informs the sender that all frames up to frame 5 have been accepted. The sender will transmit frames once it has received a receiver ready frame from the receiver.

Figure 6.9 shows an example of data flow in only the one direction.

With ABM both stations can transmit and receive data. Thus each frame sent contains receive and send counter values. When stations send information frames the previously received frames can be acknowledged, or rejected, by piggy-backing the receive counter value. In Figure 6.10, node A sends three information frames with $I[N(S)=0,N(R)=0]$, $I[N(S)=1, N(R)=0]$, and $I[N(S)=2,N(R)=0]$. The last frame informs node B that node A expects to receive frame 0 next. Node B then sends frame 0 and acknowledges the reception of all frames up to, and including frame 2 with $I[N(S)=0,N(R)=3]$, and so on.

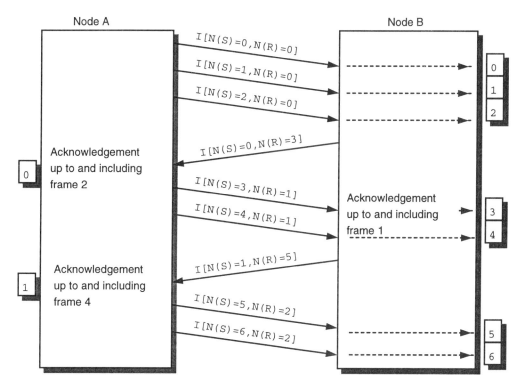

Figure 6.10   Example flow with piggy-backed acknowledgement

6.5 DERIVATIVES OF HDLC

There are many derivatives of HDLC, including:

- LAPB (link access procedure balanced) is used in X.25 packet switched networks;
- LAPM (link access procedure for modems) is used in error correction

modems;

• LLC (logical link control) is used in Ethernet and Token Ring networks;

• LAPD (link access procedure D-channel) is used in Integrated Services Digital Networks (ISDNs).

6.6 EXERCISE

6.1   Which of the following is the start and end bit sequence in an HDLC:

    A   00000000
    B   01010101
    C   01111110
    D   10000001

6.2   Which of the following has had a bit inserted by bit stuffing:

    A   11111010
    B   11101111
    C   11110111
    D   11111100

6.3   The bit sequence 0111111001111001 has had a bit inserted, in which of the bits:

    A   the 8th bit
    B   the 9th bit
    C   the 15th bit
    D   the 16th bit

6.4   In an HDLC frame how many bits does the FCS field use:

    A   8
    B   12
    C   16
    D   32

6.5 In HDLC, which of the following defines the normal response mode:

		✓

A the secondary station can only transmit when instructed by the primary

B the secondary is allowed to transmit whenever it wants

C both stations can act as a primary or a secondary

D both stations act as a primaries

6.6 In HDLC, which of the following defines the asynchronous balanced response mode:

		✓

A the secondary station can only transmit when instructed by the primary

B the secondary is allowed to transmit whenever it wants

C both stations can act as a primary or a secondary

D both stations act as a primaries

6.7 In HDLC, which of the following defines the asynchronous balanced mode.

		✓

A the secondary station can only transmit when instructed by the primary

B the secondary is allowed to transmit whenever it wants

C both stations can act as a primary or a secondary

D both stations act as a primaries

6.8 In HDLC, which type of frames sole purpose is to provide flow control information:

		✓

A information frame

B supervisory frame

C unnumbered frame

D command frame

6.7 TUTORIAL

6.9 Explain how the control field in HDLC is used to identify different types of frames. Discuss how these frames are used to transmit data.

6.10 Explain how the station can identify if an 8- or a 16-bit control field is being used.

6.11 Explain how stations use the P/F bit in the control field.

6.12 Explain the technique of piggy-back acknowledgement and discuss the mode it is used in. Sketch an example sequence between two stations.

6.13 (a) How many frames, with an 8-bit control field, can be sent before an acknowledgement is required.

(b) How many can be sent using a 16-bit control field.

6.14 With reference to Chapter 3 discuss the similarities between the LLC layer and HDLC.

6.15 The following HDLC message has been sent:

First bit sent
↓
01111000110111100111111001111101000110000001
00001XXXXXX::::::::XXXXXXXXX01001000110110
10100100111110110001111101111110100010000100
11111

Determine the bit pattern of the 8-bit address field, the 8-bit control field and FCS field. Note that the X's represent don't cares.

7

WAN: X.25 packet switching

7.1 INTRODUCTION

A wide area network (WAN) connects one node to another over relatively large distances via an arbitrary graph of switching nodes. For the transmission of digital data, then data is either sent through a public data network (PDN) or through dedicated company connections.

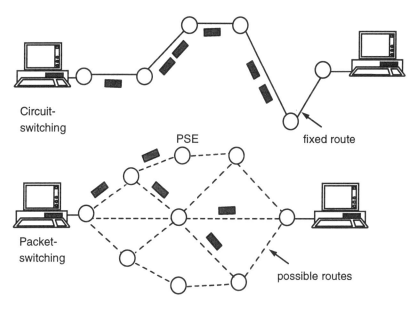

Figure 7.1 Circuit- and packet-switching

As shown in Figure 7.1, there are two main types of connection over the public telephone network, circuit-switching and packet-switching.

With circuit switched, a physical, or a reserved multiplexed, connection exists between two nodes, a typical example is the public-switched telephone network (PSTN). The connection must be made before transferring any data. In the past this connection took a relatively long time to set-up (typically over 10 seconds), but with the increase in digital exchanges it has reduced to less than a second. The usage of digital exchanges has also allowed the transmission of digital data, over PSTNs, at rates of 64 kbps and greater. This type of network is known as a circuit-switched digital network (CSDN). Its main disadvantage is that a permanent connection is set-up between the nodes. This is wasteful in time and can be costly. Another disadvantage is that the transmitting and receiving nodes must be operating at the same speed. A CSDN, also, does not perform any error detection or flow control.

Packet-switching involves segmenting data into packets that propagate within a digital network. They either follow a pre-determined route or are routed individually to the receiving node via packet-switched exchanges (PSE) or routers. These examine the destination addresses and based on an internal routing directory pass it to the next PSE on the route. As with circuit-switching, data can propagate over a fixed route. This differs from circuit-switching in that the path is not an actual physical circuit (or a reserved multiplexed channel). As it is not a physical circuit it is normally defined as a virtual circuit. This virtual circuit is less wasteful on channel resources as other data can be sent when there are gaps in the data flow.

Table 7.1 gives a comparison of the two types.

Table 7.1   Comparison of switching techniques

	Circuit-switching	*Packet-switching*
Investment in equipment	Minimal as it uses existing connections	Expensive for initial investment
Error and flow control	None, this must be supplied by the end users	Yes, using the FCS in the data link layer
Simultaneous transmissions and connections	No	Yes, nodes can communicate with many nodes at the same time and over many different routes
Allows for data to be sent without first setting up a connection	No	Yes, using datagrams
Response time	once the link is set-up it provides a good reliable connection with little propagation delay	Response time depends on the size of the data packets and the traffic within the network

7.2 PACKET-SWITCHING AND THE OSI MODEL

The CCITT developed the X.25 standard for packet switching and it fits-in well with the OSI model. In a packet-switched network the physical layer is normally defined by the X.21 standard and the data link layer by a derivative HDLC, known as LAPB. The network, or packet, level is defined by X.25.

7.2.1 The physical layer (X.21)

The CCITT recommendation X.21 defines the physical interface between a node (the DTE) and the network connection (the DCE). Figure 7.2 shows the connections between the node and the network connection.

A second standard, known as X.21 (bis), has also been defined and is similar to the RS-232/V.24 standards. This allows RS-232 equipment to directly connect to the network.

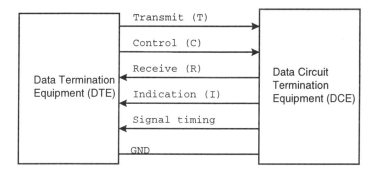

Figure 7.2 X.21 connections

The Transmit (T) line sends data from the DTE to the DCE and the Receive (R) sends data from the DCE to the DTE. A DCE controls the Indicate (I) line to indicate that it is ready to receive data. The DTE controls the Control (C) line to request to the DCE that it is ready to send data.

Figure 7.3 shows a simplified flow control between a sending DTE and a receiving DCE. With reference to the state numbers in the diagram the sequence of operations is as follows:

1   Initially, the Transmit (T) and Receive (R) lines are high to indicate that the DTE and the DCE are active and ready to communicate, respectively.

2   When the DTE wishes to transmit data it first sets the Control (C) line low (ON). At the same time it sets the Transmit (T) line low.

3   When the DCE accepts the data transfer it sets the Indicate (I) line low.

4-12 Data is transmitted on the Transmit (T) line and, in some modes, it is echoed back on the Receive (R) line.

12   When the DTE finishes transmitting data it sets the Control (C) line high (OFF).

13   The DCE responds to the Control (C) line going high by setting the Indication (I) line high.

14   The DCE sets the Receive (R) line high to indicate that it is active and ready to communicate.

15   The DTE sets the Transmit (T) line high to indicate that it is active and ready to communicate.

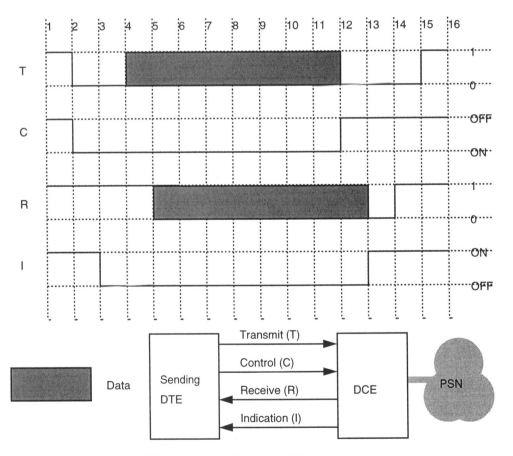

Figure 7.3   Example X.21 signals

7.2.2 Data link layer (LAPB)

The data link layer provides a reliable method of transferring packets between the DTE and the local PSE. Frames sent contain no information on the addressing of the remote node, this information is contained within the packet. The standard, known as the Link Access Procedure Version B is based on HDLC. It uses ABM (asynchronous balance mode) where both the DTE and the PSE can initiate commands and responses at any time.

7.2.3 Network (packet) layer

The packet layer is equivalent to the network layer in the OSI model. Its main purpose is to route data over a network.

7.3 X.25 PACKETS

X.25 packets contain a header and either control information and/or data. The LAPB frame envelops the packet and physical layer transmits it. Figure 7.4 shows the format of the transmitted frame.

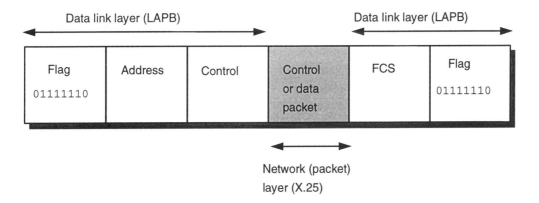

Figure 7.4 Transmitted frame

7.3.1 Packet headers

Packet headers are 3 bytes long. Figure 7.5 shows a packet represented both as a bit stream and arranged in groups of 8 bits. The first two bytes of the header contain the group format identifier (GFI), the logical group number (LGN) and the logical channel number (LCN). The third byte identifies the packet type.

The GFI number is a 4-bit binary value of QdYY, where the Q bit is the qualifier bit and the D bit is the delivery confirmation bit. The d bit requests an acknowledgement from the remote node, this is discussed in more detail in section 7.4.3. The YY bits indicate the range of packet sequence numbers. If they are 01 then packets are numbered from 0 to 7 (modulo-8), or if they are 10 then packets are numbered from 0 to 127 (modulo-128). This packet sequencing is similar to the method that HDLC uses to provide confirmation of received frames. As LAPB (the HDLC-derivative) provides reliable data link error control, the sequencing of packets is mainly used as flow control rather than for error control.

The LGN and LCN together define a 12-bit virtual circuit identifier (VCI). This allows packets to find logical routes though the packet switched network. For example all packets could take the same route or each group of packets could find different routes. This will be discussed in more detail in section 7.5.

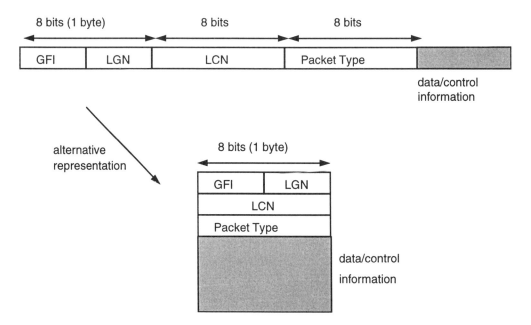

Figure 7.5   Packet header

7.3.2 Packet types

The third byte of the packet header defines the packet type, Table 7.2 lists some of these. A 0 in the eighth bit position identifies that it is a data packet while a 1 marks it as a flow control or a call set-up packet.

Table 7.2   The main packet types

Packet type	Identifier	Description
Data Packet	RRRMSSS0	The Data packet is sent with a send sequence number (SSS) and a receive sequence number (RRR).
Call Request/ Incoming Call	00001011	If a node sends this packet to the network it is a Call Request packet, if the node receives it, it is an Incoming Call packet.
Call Accepted/ Call Confirmation	00001111	If a node sends this packet to the network it is a Call Accepted packet, if a node receives it, it is a Call Confirmation.
Receive Ready	RRR00001	The Receive Ready packet is sent from a node to inform the other node that it is ready to receive data. It also informs the other node that the next data packet it expects to receive should have sequence number RRR.
Receive Not Ready	RRR00101	The Receive Not Ready packet is sent from a node to inform the other node that it is not ready to receive data. It also informs the other node that the next data packet it expects to receive should have sequence number RRR.
Reject	RRR01001	The Reject packet is sent from a node to inform the other node that it rejects packet number RRR. All other packets before this are acknowledged.
Clear Request/ Clear Indication	00010011	If the calling node sends this packet to the network it is a Clear Request packet, if the called node receives it is a Clear Indication packet.
Clear Confirm/ Clear Confirm	00010111	If the called node sends this packet to the network it is a Clear Confirm packet, if the calling node receives it is a Clear Confirm.

The call set-up and clearing packets have differing definitions depending on whether they are sent or received. For example if the calling node sends a packet type of 00001011 then it is a Call Request packet, if it is received it is interpreted as an Incoming Call packet.

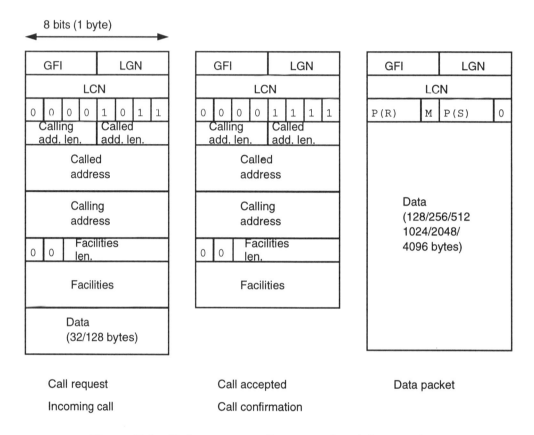

Figure 7.6   Call request, call accepted and data packets

Figure 7.6 shows the format of the Call Request/ Incoming Call, Call Accepted/ Call Confirmation and the Data packets. With the Call Request/ Incoming Call and the Call Accepted/ Call Confirmation packets the fourth byte of the packet contains two 4-bit numbers which define the number of bytes in the calling and called address. After this byte, the called and the calling addresses are sent. Following this the next byte defines the number of bytes in the facilities field. The facilities field enables selected operational parameters to be negotiated when a call is being set up, these include:

- the data packet size (typically, 128 bytes);
- number of packets to be received before an acknowledgement is required (typically, two);
- data throughput, in bytes per second;
- reverse charging;
- usage of extended sequence numbers.

A data packet contains the standard packet header followed by a byte that contains the send and receive sequence number. The M bit identifies that there is more data to be sent to complete the message. Notice that the data packet does not contain either the calling or the called addresses. This is because once the connection is made then the VCI label identifies the path between the called and the calling node.

The P(R) variable is the sequence number of the packet that the sending node expects to receive next, and P(S) is the sequence number of the current packet. With modulo-8 sequencing, the packets are numbered from 0 to 7. The first packet sent is 0, the next is 1 and so on until the eighth packet that is numbered 7. The next is then numbered as 0 and so on. With modulo-128 sequencing, the packets are numbered 0 to 127.

The data size can be 128, 256, 512, 1024, 2048 or 4096 bytes, although its size is normally limited, by the public-carrier packet-switched network, to 128. This achieves a reasonable response time.

Figure 7.7 shows the format of the Receive Ready (RR), Receive Not Ready (RNR), Reject (REJ), Clear Confirmation and Clear Request packets. The RR, RNR and REJ packets contain a receive sequence number. This is the sequence number of the packet that the receiving node expects to receive next.

7.4 X.25 PACKET FLOW

The three types of packets are:

- call set-up and clearing - Call Request, Incoming Call, Clear Request, Clear Indication and Clear Confirmation;
- data packets;
- flow control - Receive Ready (RR), Receive Not Ready (RNR) and Reject (REJ).

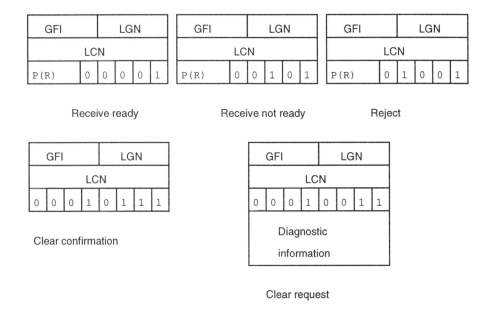

Figure 7.7   RR, RNR, REJ and Clear Confirm and Request packets

7.4.1 Call set-up and clearing

Figure 7.8 shows a typical data transfer. Initially the calling node (Node A) sends a Call Request packet (P[Call_request]) to the network. When this propagates through the packet-switched network the receiving node (Node B) receives it as an Incoming Call packet (P[Incoming_call]). When Node B accepts the call it sends a Call Accepted packet (P[Incoming_call]), which propagates through the network and Node A receives it as a Call Confirmation (P[Call_confirmation]).

The call initialization sets up a virtual circuit between the nodes and sequenced data packets and flow control information can now flow between the nodes.

To clear the connection Node A sends a Clear Request packet (P[Clear_request]) to the network. When this propagates through the network, Node B receives it as a Clear Indication packet (P[Clear_indication). When Node B accepts that the call is to be cleared then it sends a Clear Confirmation packet (P[Clear_confirm]). This propagates through the network and Node A receives it as a Clear Confirm (P[Call_confirmation]).

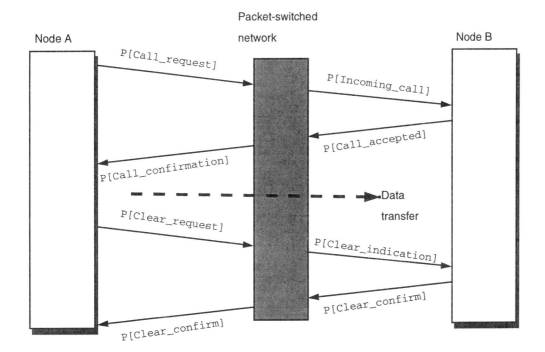

Figure 7.8   Call set-up and clearing

7.4.2 Data transmission and flow control

After a virtual circuit has been set-up then sequenced data packets and flow control information can flow between the nodes.

A data packet contains sends two sequence numbers P[D, N(R), N(S)]. For 3-bit sequence numbers, N(S) and N(R) range from 0 to 7. N(S) is the sequence number of the data packet and N(R) is the sequence number of the packet that the sending node expects to receive next.

Figure 7.9 shows an example conversation between a sending node (Node A) and a receiving node (Node B). The flow control window has been set at 3. This window defines the number of packets that can be sent before the receiver must send an acknowledgement. Initially, in the example, three information frames are sent, numbered 2, 3 and 4 (P[D, N(R)=0, N(S)=2], P[D, N(R)=0, N(S)=3] and P[D, N(R)=0, N(S)=4]). The window is set to 3 thus Node B must send an acknowledgement for the packets it has received. It does this by sending a Receive Ready packet (P[RR, N(R)=5]). This informs Node A that

Node B expects to receive packet number 5 next. This acknowledges all frames before, and including, frame 4.

In the example in Figure 7.9 an error has occurred in the reception of frame 5. The recipient informs the sender by sending a Reject packet (P[REJ, N(R)=5]). After the sender receives this it re-sends each frame after, and including, frame 5.

If a node does not wish to communicate, at the present, it sends a Receive Not Ready packet. For example P[RNR, N(R)=5] tells the transmitter to stop sending data, at the present. It also informs the sender that all frames up to frame 5 have been accepted. The sender will transmitting frames only once it has received a Receive Ready packet from the receiver.

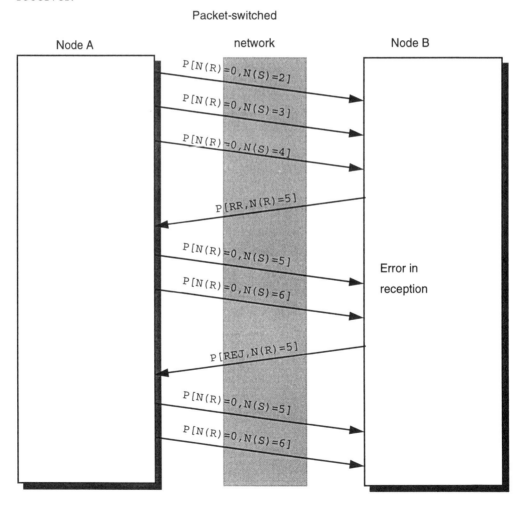

Figure 7.9   Example flow

Figure 7.8 shows an example of data flow in only the one direction. With X.25 both stations can transmit and receive data. Thus each packet sent contains receive and send counter values. When nodes send data packets the previously received frames can be acknowledged, or rejected, by piggy-backing the receive counter value. In Figure 7.9, Node A sends 3 data packets with `P[D, N(R)=0, N(S)=0]`, `P[D, N(R)=0, N(S)=1]`, and `P[D, N(R)=0, N(S)=2]`. The last data packet informs Node B that Node A expects to receive data packet 0 next. Node B then sends data packet 0 and acknowledges the reception of all frames up to, and including frame 2 with `P[D, N(R)=3, N(S)=0]`, and so on.

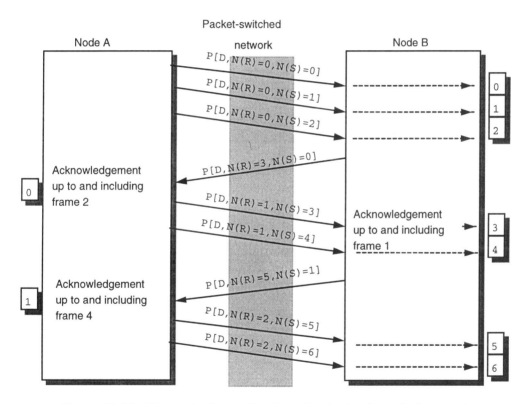

Figure 7.10 Example flow with piggy-backed acknowledgement

7.4.3 The delivery bit

The delivery bit (d) identifies which connection should respond with an acknowledgement. If it is not set then the local network connection sends an acknowledgement for data packets. If it is set then the remote nodes sends the acknowledgement. The latter was the case in the example given in Figure 7.9. An example is given in Figure 7.11. The number of packets before an acknowledgement is set by the window. When the d-bit is not set then the window does not have any significance as the network connection returns back all packets sent.

7.5 PACKET SWITCHING ROUTING

There are three main types of routing used in X.25, these are:

- permanent virtual call;
- virtual call;
- datagram.

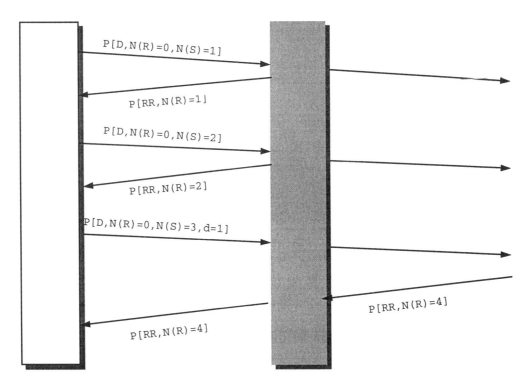

Figure 7.11   Usage of the D-bit

A virtual call sets up a route for the two-way flow of packets between two specific nodes, whereas a datagram is sent into the network without first establishing a route. A datagram is analogous to a letter sent by the post, where a letter is addressed and sent without first finding out if the letter will be received. The virtual call is analogous to a telephone call where a direct connection is made before the call is initiated.

A datagram is normally only used where there is a small amount of data in a few packets, whereas a virtual call is set-up where there are relatively large amounts of data to be sent in many packets. With a datagram there is no need to initiate the call set-up procedures, as previously shown in Figure 7.8. The call set-up and clearing packets (Call Request, Call Indication, and so on) are only used when a virtual circuit is used. An example of a datagram might be to transmit an electronic mail message as there is no need to establish a virtual circuit before the message is sent.

Once a virtual circuit is set-up then it is used until all the data has been transferred. A new conversation establishes a new virtual circuit. In some applications, though, a reliable permanent circuit is required, this is described as a permanent virtual call where two nodes have a permanent virtual connection. There is no need, in this case, to set-up a connection as the virtual circuit between the two nodes is dedicated to them.

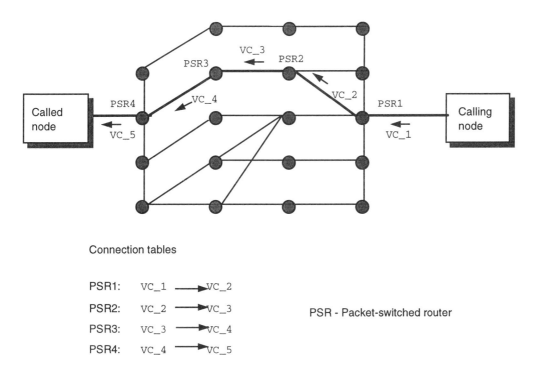

Figure 7.12   Virtual call set-up

When a calling node wishes to communicate through the network it makes contact with the called node and negotiates such things as the packet size, reverse charging and maximum data throughput. The window size is the number of packets that are sent before an acknowledgement must be sent back. Maximum data throughput is the maximum number of bytes that can be sent per second. These flow control parameters are contained in the facilities field of the Call Request packet. If the called node accepts them then a connection is made.

The network computes a route based on the specified parameters and determines which links on each part of the route best supports the requested flow control parameters. It sets up request to all the packet routing nodes on the path en-route to the destination node. Figure 7.12 shows an example route between packet-switched routers (PSR) from a calling node to a destination node. The route selected is PSR1 $\rightarrow$ PSR2 $\rightarrow$ PSR3 $\rightarrow$ PSR4. Each of the router selects an unused VCI label on their respective links and reserves it for the virtual circuit in their connection lookup tables.

For example PSR1 could use VC_2 (for example the VCI could have a value of 17), this can be sent to PSR2. PSR2 in turn picks VC_3 and associates it with VC_2 in its connection table. It then forwards VC_3 to PSR3. PSR3 selects VC_4 and associates it with VC_3. It then forwards VC_4 to PSR4. PSR4 selects VC_5 and associates it with VC_4. If the called node accepts the call then it sends back a Call Accepted packet back over the virtual circuit. Each of the nodes on way back to the calling node assigns a new VCI number. Thus the acknowledgement passes back from PSR4 to PSR3, PSR3 to PSR2 and so on to the calling node to confirm that the connection has been established. Data packets can then be transmitted between the two nodes. As has been previously discussed, there is no need to transmit the calling or called addresses with the packet as the source and destination is identified using the VCI label. When the connection is terminated the VCI labels assigned to the communications can be used for other connections.

If a single virtual circuit is set-up then packets are always delivered in the order these were transmitted. This is because packets cannot take alternative routes to the destination. Even if the packets are buffered within a node they will still be transmitted in the correct sequence.

7.6 LOGICAL CHANNELS

The VCI label contains a 4-bit logical group number (LGN) and a 16-bit logical channel number (LCN) to define the VCI. There can be 16 groups and within each group there can be 256 different channels. This allows for a node to communicate with several nodes simultaneously. Figure 7.13 shows an example of a node communicating with four nodes over four channels. Node A is communicating with Node B, C, D and E. The route for Node A to Node E is through routers A, B, C, G, and I. For Node A to Node D it is through routers A, F, G and H.

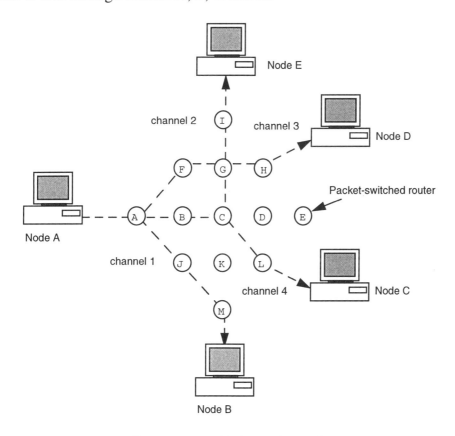

Figure 7.13 Multiple channels

7.7 X.25 NODE ADDRESSING

Nodes on an X.25 network have an individual NSAP (network service access point) address. Since these nodes operate globally over international

networks the addresses must be assigned a globally unique network address. The network on which the node is connected is usually a country-
wide network. Each of these packet-switched public digital networks are
known as a subnet.

The definition of the addresses is defined either in pure binary or a binary-coded decimal (BCD) digits. For example, if the network address is
defined in BCD then the binary address 0011 0110 0001 0111
1001 0011 corresponds to 361792. The NSAP address is made up of
up-to 40 decimal digits or 20 bytes (as one BCD digit is represented by 4
bits). The calling and called address length are defined within the X.25
packet.

The NSAP address is made up of parts, the initial domain part (IDP)
and the domain specific part (DSP), as illustrated in Figure 7.14. An IDP
is made up of two sub-parts, the authority and format identifier (AFI), and
the initial domain identifier (IDI). As several authorities can grant NSAP
addresses, the AFI field contains 2 BCD digits which identify the granting
authority and the format of the rest of the address field.

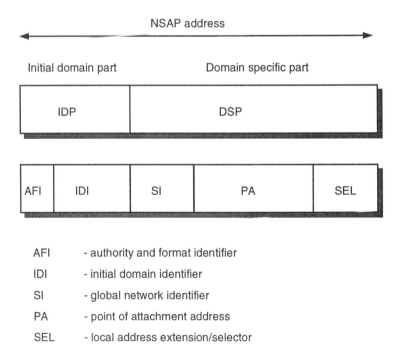

Figure 7.14   NSAP address

For example, if the AFI value is 36 then the granting authority is the

CCITT and the format is defined in the X.121 recommendation. The resulting address is:

`36XXXXXXXXXXXXXXXXXXXYYYYYYYYYYYYYYYYYYYYYYYY`

where `XX..XXX` is the IDI part (14 digits) and `YY..YYY` is the DSP part (24 digits). This gives a total of 40 digits.

If the AFI value is `38` then the granting authority is the ISO and the format is defined by the ISO-assigned country codes, or ISO DCC. The resulting address is:

`38XXXYYYYYYYYYYYYYYYYYYYYYYYYYYYYYYYYYYY`

where `XX..XXX` is the IDI part (3 digits) and `YY..YYY` is the DSP part (35 digits). With ISO address the IDI portion is assigned by the country the network is resident.

After the initial domain is defined then the DSP part defines a smaller and smaller subnetwork within the domain. Figure 7.15 shows an example addressing structure. The SEL part defines the local node with at the point of attachment of the packet switched network.

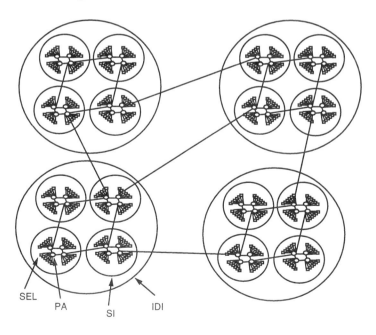

Figure 7.15: NSAP addressing structure

7.8 EXERCISE

7.1   Which standards organization defined the X.25 standard:

    A   IEEE
    B   CCITT
    C   ANSI
    D   ISO

7.2   The standard for the physical layer, in packet switching, is:

    A   X.21
    B   X.12
    C   X.25
    D   RS-232C

7.3   The main functional difference between circuit-switched digital networks (CSDN) and packet-switched digital network (PSDN) is:

    A   that more data can be transmitted
    B   that they are more reliable
    C   that they allow different types of data to be transmitted
    D   that a CSDN establishes a permanent connection between two nodes where PSDN uses a virtual circuit

7.4   In X.21, which signal line does the DTE control:

    A   Control
    B   Receive
    C   Indication
    D   Clear

7.5   Which standard is used for the data link layer for the connection between a node and a packet-switched network:

A X.21
B RS-232
C X.25
D LAPB

7.6 When using modulo-8 sequence numbers, what is the range of the flow window:

A 1–2
B 1–7
C 1–8
D 1–16

7.7 When using modulo-128 sequence numbers, what is the range of the flow window:

A 1–8
B 1–126
C 1–127
D 1–256

7.8 Typically, what is the size of flow window:

A 1
B 2
C 3
D 8

7.9 Typically, how much data is sent within a data packet

A 128 bytes
B 256 bytes
C 512 bytes
D 4096 bytes

7.9 TUTORIAL

7.10   Explain the difference between a virtual call (circuit) and a datagram.

7.11   Discuss the usage of the more-data bit (M) and the delivery confirmation bit (d).

7.12   Explain why, with a virtual circuit, a node does not need to send its own, and the destination's address, when sending data and flow control.

7.13   Explain the use of piggy-back acknowledgement and show an example flow of data packets.

7.14   Explain why packet-switching is more reliable than circuit-switching when transmitting data.

7.15   Explain why the propagation delay of a circuit-switched connection is constant but in a packet-switched switched network it is dependent upon the network traffic.

8

Asynchronous communications: RS-232

8.1 INTRODUCTION

In 1962 the Electronics Industries Association (EIA) introduced the RS-232C standard to define the interface between Data Terminal Equipment (DTE) and Data Circuit-termination Equipment (DCE). Since then the growing use of PCs has ensured that RS-232C became an industry standard for all low-cost serial interfaces between the DTE (the computer) and the peripheral. Typical implementations include serial mouse, printers, plotters, accessing remote instrumentation, scanners, digitizers and modems. In 1978 the EIA updated the original standard with a new one called RS-232D.

RS-232C is a standard which covers most of the physical layer and also part of the data link layer.

8.2 COMMUNICATIONS TECHNOLOGY

This section discusses the terminology used in RS-232 communications.

8.2.1 Connecting two nodes

The three main modes of communications are shown in Figure 8.1, these are:

- simplex communications – one-way communications in which data can only flow from one node to the other;
- half-duplex communications – two-way communications but only one node can communicate at a time;
- full-duplex communications – two-way communications in which both devices can communicate simultaneously.

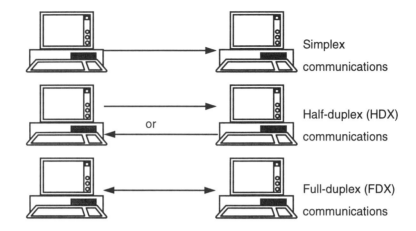

Figure 8.1: Serial Communications

8.2.2 Terminal and communications equipment

The end of the data communication channel is known as the Data Terminal Equipment (DTE), such as a computer or a fax. Data Circuit-termination Equipment (DCE) is the communications channel between two DTEs, such as a modem or connecting cable. Figure 8.2 shows the differing definitions of communications equipment.

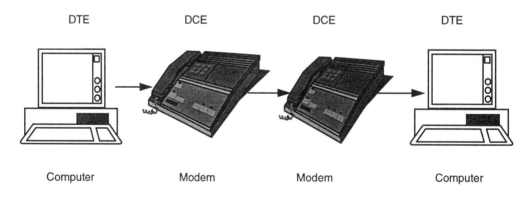

Figure 8.2: A DTE and a DCE

8.3 ASCII CHARACTER CODES

Data communication is the transmission of digital information from a source to a destination. Normally this information is sent in a coded form. These could take the form of ASCII characters, microprocessor machine codes, control words, and so on.

The three most common character sets currently in use are Baudot code, American Standards Code for Information Interchange (ASCII) code and the Extended Binary-Coded Decimal Interchange Code (EBCDIC). RS-232 uses the ASCII character set.

ASCII is a standard international character alphabet where character is defined as a 7-bit character code, with 32 control and 96 printable characters. The bits are identified as b_0 (the lsb) to b_6 (the msb). An eighth bit (b_7) can be used as a parity bit to give a degree of error detection. For even parity the parity bit is added so that it evens up the number of 1's. For odd parity it makes the number of 1's odd.

In RS-232 communications the least significant bit of the ASCII character is sent before all the other bits. The last bit sent is the parity bit. Figure 8.3 shows an example of the message 'Fred' sent using odd parity.

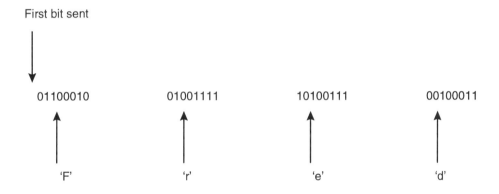

Figure 8.3: Example message showing first bit sent

📄 **Program 8.1**

```
#include <stdio.h>

int  main(void)
{
int i;
   for (i=0;i<128;i++)
   {
     printf("<%3d%2c>",i,i);
     if (!(i % 6)) printf("\n");
   }
   return(0);
}
```

When displaying ASCII characters to the screen there are many unseen characters which are not displayed. These include carriage return (CR, 0Dh), space (SP, 20h) horizontal tab (HT, 09h), and so on. Program 8.1 displays the ASCII characters from 0 to 127.

Test run 8.1 shows a sample run. Notice that the ASCII character 10 (NL) has generated a new line, character 8 (BS) caused a back-space and character 9 (HT) has generated a horizontal tab. Also, when running the program the computers speaker sounds when printing character 7 (BEL).

```
Test run 8.1
<   0   >
<   1   _> <   2   _> <   3   _> <   4   _> <   5   _> <   6   _>
<   7   > <   8   > <   9      > < 10
> < 11        _> < 12
>
> < 14   _> < 15   ⌐> < 16   _> < 17   _> < 18   _>
< 19   _> < 20   ¶> < 21   §> < 22   _> < 23   _> < 24   _>
< 25   _> < 26   > < 27   _> < 28   _> < 29   _> < 30   _>
< 31   _> < 32   > < 33   !> < 34   "> < 35   #> < 36   $>
< 37   %> < 38   &> < 39   '> < 40   (> < 41   )> < 42   *>
< 43   +> < 44   ,> < 45   -> < 46   .> < 47   /> < 48   0>
< 49   1> < 50   2> < 51   3> < 52   4> < 53   5> < 54   6>
< 55   7> < 56   8> < 57   9> < 58   :> < 59   ;> < 60   <>
< 61   => < 62   >> < 63   ?> < 64   @> < 65   A> < 66   B>
< 67   C> < 68   D> < 69   E> < 70   F> < 71   G> < 72   H>
< 73   I> < 74   J> < 75   K> < 76   L> < 77   M> < 78   N>
< 79   O> < 80   P> < 81   Q> < 82   R> < 83   S> < 84   T>
< 85   U> < 86   V> < 87   W> < 88   X> < 89   Y> < 90   Z>
< 91   [> < 92   \> < 93   ]> < 94   ^> < 95   _> < 96   `>
< 97   a> < 98   b> < 99   c> <100   d> <101   e> <102   f>
<103   q> <104   h> <105   i> <106   j> <107   k> <108   l>
<109   m> <110   n> <111   o> <112   p> <113   q> <114   r>
<115   s> <116   t> <117   u> <118   v> <119   w> <120   x>
<121   y> <122   z> <123   {> <124   |> <125   }> <126   ~>
<127   @>
```

Format codes and control characters are non-printing characters. These codes are decimal 0 to 31. Table 8.1 gives a listing of these.

8.3.1 Format codes

Backspace (BS)

On displays the backspace character (000 1000) erases the previous character sent. Normally the backspace key on a keyboard generates a ^H (Cntrl-H) character (although some keyboards return a ^?).

Horizontal Tab (HT)

The horizontal tab character (000 1001) feeds the current display cursor forward by one tab spacing. Tab settings are normally set by the computer or by a software package. Most keyboards have a TAB key which return a ^I character.

Line Feed (LF), Carriage Return (CR)

The carriage return character (000 1101) returns the cursor display position to the beginning of a line. A line feed (000 1010) forces a new line and moves the cursor position down one position. On Unix systems a new-line character is normally defined by the CR/LF sequence whereas on PC systems it is defined by the line feed character.

Table 8.1 Non-printing ASCII characters

ASCII character	ASCII decimal	Binary code	Hex code	Control character	Function
NUL	0	000 0000	00	^@	Null
SOH	1	000 0001	01	^A	Start of Heading
STX	2	000 0010	02	^B	Start of Text
ETX	3	000 0011	03	^C	End of Text
EOT	4	000 0100	04	^D	End of Transmission
ENQ	5	000 0101	05	^E	Enquiry
ACK	6	000 0110	06	^F	Acknowledge
BEL	7	000 0111	07	^G	Bell
BS	8	000 1000	08	^H	Backspace
HT	9	000 1001	09	^I	Horizontal Tab
LF	10	000 1010	0A	^J	Line Feed
VT	11	000 1011	0B	^K	Vertical Tab
FF	12	000 1100	0C	^L	Form Feed
CR	13	000 1101	0D	^M	Carriage Return
SO	14	000 1110	0E	^N	Shift Out
SI	15	000 1111	0F	^O	Shift In
DLE	16	001 0000	10	^P	Data Line Escape
DC1	17	001 0001	11	^Q	Device Control 1
DC2	18	001 0010	12	^R	Device Control 2
DC3	19	001 0011	13	^S	Device Control 3
DC4	20	001 0100	14	^T	Device Control 4
NAK	21	001 0101	15	^U	Negative Acknowledge
SYN	22	001 0110	16	^V	Synchronous Idle
ETB	23	001 0111	17	^W	End of Transmit Block
CAN	24	001 1000	18	^X	Cancel
EM	25	001 1001	19	^Y	End of Medium
SUB	26	001 1010	1A	^Z	Substitute
ESC	27	001 1011	1B	^[, ESC	Escape
FS	28	001 1100	1C	^\	File Separator
GS	29	001 1101	1D	^]	Group Separator
RS	30	001 1110	1E	^6	Record Separator
US	31	001 1111	1F	^_	Unit Separator

Form Feed (FF)

The form feed character (000 1100) causes a line printer to feed to the next page. If it is sent to a display it moves the display cursor one position to the right (the opposite effect to the backspace character).

Vertical Tab (VT)

When sent to a printer the vertical tab (000 1011) character causes the current position to move to the next programmed vertical tab space. If it is sent to a display it moves the current cursor position one line upwards.

Normally, a vertical tab is defined by a ^K keystroke.

Many keyboards return the following characters when the arrowkeys are pressed. Some displays also interpret these characters as movement commands.

Cursor left	^H
Cursor down	^J
Cursor up	^K
Cursor right	^L

8.3.2 Communication-control characters

End of Text (EXT)
Most computer systems use the end of text character (000 0100) to interrupt a process (^C).

End of Transmission (EOT)
Unix systems use the end-of-transmission character (000 0101) to signal that the user has finished entering data. It is also as an end-of-file character (^D).

Substitute (SUB)
PC-based DOS systems use the SUB character (001 1010) to define an end of file (EOF).

8.3.3 Other codes

Null (NUL)
Often the null character (000 0000) pads the beginning of transmitted characters and also delimits a string of characters. For example, the word 'FRED' can be sent as 'F', 'R', 'E', 'D' , NUL. The control character used to represent the null character is ^@.

Bell (BEL)
When sent to a device with a speaker the bell character (000 0111) generates a tone. The control character used to represent the bell is ^G.

Device control (DC1-DC4)
DC1-DC4 control the operation of displays and printers. A receiver uses the DC1 (^Q) character to inform the transmitter to stop transmitting. The DC3 (^S) character inform the transmitter to start transmitting again. This

type of communications is known as software handshaking.

Esc (escape)
Used by some packages to escape from menu options, etc.

Del (delete)
The delete character (111 1111) erases the character at the current cursor position.

8.3.4 Printable character set

The printable characters are all displayable. They include all upper- and lower-case letters ('a'–'z', 'A'–'Z'), numerical characters ('0'–'9') and other characters such as SPACE, '!' and '#'. The codes for upper- and lower-case letters only differ by one bit position (the b_5 bit), for example:

'A'	10**0** 0001
'a'	11**0** 0001
'B'	10**0** 0010
'b'	11**0** 0010
'Z'	10**1** 1010
'z'	11**1** 1010

Program 8.2 reads a file and displays the integer values of the characters in a file and Test run 7.2 shows a sample run of the program with the file FILE.TXT (as shown in File listing 8.1). The first character in the file, 'T', is ASCII decimal 84, the next, 'h' is decimal 104, and so on. Notice that the new line character is decimal 10, as the program was run on a PC-based system using DOS.

📄 **Program 8.2**
```c
#include <stdio.h>
int  main(void)
{
FILE    *in;
char    ch, fname[BUFSIZ];
  printf("\nEnter a file to read >> ");
  gets(fname);

  if ( (in=fopen(fname,"r"))==NULL)
    printf("Cannot open file <%s>\n",fname);
  else
  {
    do
    {
```

```
        fscanf(in,"%c",&ch);
        printf("%d ",ch);
    } while (!feof(in));
    fclose(in);
  }
  return(0);
}
```

⌨ Test run 7.2

```
Enter a file to read >> file.txt
84 104 105 115 32 105 115 32 97 110 32 101 120 97 109 112 108 101 10 111
102 32 97 110 32 65 83 67 73 10 102 105 108 101 46 10
```

📖 File listing 7.1: Listing of FILE.TXT

```
This is an example
of an ASCII
file.
```

8.4 ELECTRICAL CHARACTERISTICS

The electrical characteristics of RS-232 corresponds to the physical layer of the OSI model.

8.4.1 Line voltages

The electrical characteristics of RS-232 defines the minimum and maximum voltages for a logic 1 and a 0. A logic 1 ranges from –3 V to –25 V, but will typically be around –12 V and a logical 0 ranges from +3 V to +25 V, but will typically be around +12V. Any voltage between –3 V and +3 V has an indeterminate logical state.

If no pulses are present on the line then the voltage level is equivalent to a high level, that is –12 V. If the receiver detects a 0 V voltage level then is interpreted as a line break or short circuit. Figure 8.4 shows an example transmission.

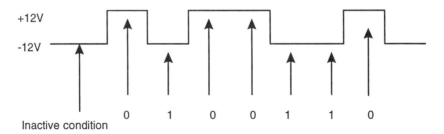

Figure 8.4   Example of RS-232 voltage levels

8.4.2 Electrical connections

The two connectors used with RS-232 are the 9- and 25-way D-type connectors. Most modern PCs use the 9-pin connector for the primary (COM1 :) and a 25-pin for a secondary serial port (COM2 :).

8.4.3 DB25S Connector

The DB25S connector is a 25-pin D-type connector which gives full RS-232 functionality, as illustrated in Figure 8.5. Table 8.2 lists the main connections. A DCE (such as a connecting cable) connector has a male outer casing with female connection pins. The DTE (such as a computer) has a female outer casing with male connecting pins. There are three main signal types: control, data and ground. Control lines are active HIGH, that is, they are HIGH when the signal is active and LOW when inactive.

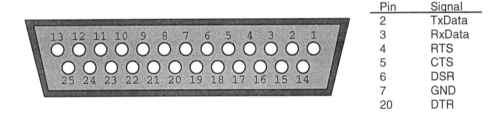

Pin	Signal
2	TxData
3	RxData
4	RTS
5	CTS
6	DSR
7	GND
20	DTR

Figure 8.5   RS-232 DB25S connector

8.4.4 DB9S connector

The DB25S connector is the standard for RS-232 connections, but as electronic equipment becomes smaller there is a need for smaller connectors. For this purpose most PCs use a reduced function 9-pin D-type connector rather than the full function 25-way D-type. As with the 25-pin connector the DCE (the terminating cable) connector has a male outer casing with female connection pins. The DTE (the computer) has a female outer casing with male connecting pins. Figure 8.6 shows the main connections.

8.4.5 PC connectors

All PCs have at least one serial communications port. The primary port is named COM1: and the secondary is COM2:, as illustrated in Figure 8.7. A primary port connector is normally a 9-pin D-type socket (or female) connector and the secondary port (if there is one) is normally a 25-pin D-type socket type. On some PCs both connectors are 9-pin types (typically on modern PCs) or are both 25-pin types (typically in older PCs). Different connector types can cause problems connecting cables. Thus a 25- to 9-pin adapter is a useful attachment, especially to connect a serial mouse to a 25-pin connector.

Table 8.2   Main pin connections used in 25-pin connector

Pin	Name	Abbreviation	Functionality
1	Frame Ground	FG	This ground is normally connects the outer sheath of the cable and to earth ground.
2	Transmit Data	TD	Data is sent from the DTE (computer or terminal) to a DCE via TD.
3	Receive Data	RD	Data is sent from the DCE to a DTE (computer or terminal) via RD.
4	Request to send	RTS	DTE sets this active when it is ready to transmit data.
5	Clear to send	CTS	DCE sets this active to inform the DTE that it is ready to receive data.
6	Data Set Ready	DSR	Similar functionality to CTS but activated by the DTE when it is ready to receive data.
7	Signal Ground	SG	All signals are referenced to the signal ground (GND).
20	Data Terminal Ready	DTR	Similar functionality to RTS but activated by the DCE when it wishes to transmit data.

Pin	Signal
2	RxData
3	TxData
4	DTR
5	GND
6	DSR
7	RTS
8	CTS

Figure 8.6   RS-232 DB9S interface

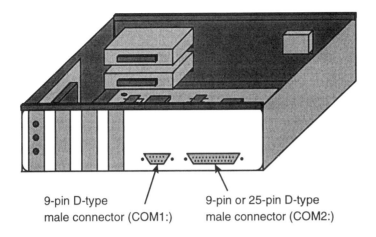

Figure 8.7 Typical PC connectors

8.5 FRAME FORMAT

RS-232 uses asynchronous communications which has a start-stop data format. Characters are transmitted one at a time with a delay in between, as shown in Figure 8.8. This delay is called the inactive time and is set at a logic level high (–12V).

The transmitter sends a start bit to inform the receiver that a character is to be sent in the following bit transmission. This start bit is always a 0. Next 5, 6 or 7 data bits are sent as a 7-bit ASCII character, followed by a parity bit and finally either 1, 1.5 or 2 stop bits. Figure 8.9 shows the frame format and an example transmission of the character 'A' using odd parity.

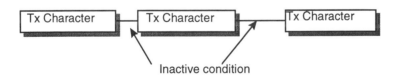

Figure 8.8 Asynchronous communications

The rate of transmission is set by the timing of a single bit. An internal clock on both the transmitter and receiver are set with the same rate. These only have to be roughly synchronized and approximately at the same rate as data is transmitted in relatively short bursts. This differs from synchronous communications which normally sends large bursts of data and thus requires accurate timing.

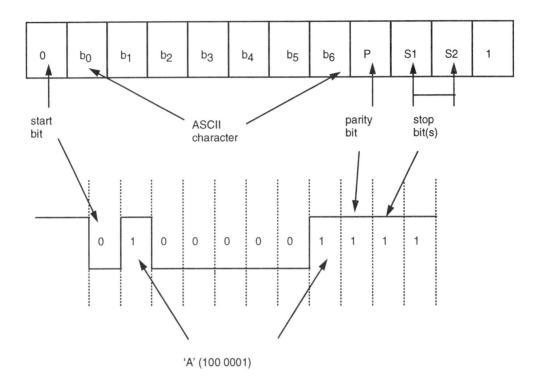

Figure 8.9   RS-232 format

Example:

An RS-232 serial data link uses 1 start bit, 7 data bits, 1 parity bit, 2 stop bits, ASCII coding and even parity. Determine the message sent from the following bit stream.

First bit sent
⇓
1111101000001011000001111111111111100000011111111
00011001111010100111111111111

ANSWER

The format of the data string sent is given next:

{idle} 11111 {start bit} 0 {'A'} 1000001 {parity bit} 0 {stop bits }
11 {start bit} 0 {'p'} 0000111 {parity bit} 1 {stop bits} 11 {idle}
11111111 {start bit} 0 {'p'} 0000111 {parity bit} 1 {stop bits} 11
{idle} 11 {start bit} 0 {'L'} 0011001 {parity bit} 1 {stop bits} 11

The message sent was thus 'AppL'.

8.5.1 Parity

Error control is data added to transmitted data in order to detect or correct
an error in transmission. RS-232 uses a simple technique known as parity
to provide a degree of error detection.

A parity bit is added to transmitted data to make the number of 1's sent
either even (even parity) or odd (odd parity). It is a simple method of error coding and only requires exclusive-OR (XOR) gates to generate the
parity bit. A simple parity generator for 5 bits is given in Figure 8.10. The
parity bit is added to the transmitted data by inserting it into the shift register at the correct bit position.

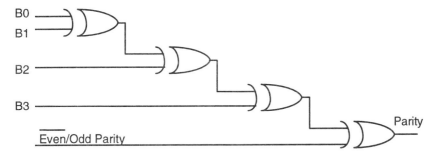

Figure 8.10   Circuit to generate parity bit using EX-OR gates

A single parity bit can only detect an odd number of errors, that is, 1, 3,
5, etc. If there are an even number of bits in error then the parity bit will
be correct and no error will be detected. This type of error coding is not
normally used on its own where there is the possibility of several bits being in error.

8.5.2 Baud rate

One of the main parameters which specify RS-232 communications is the
rate of transmission at which data is transmitted and received. It is important that the transmitter and receiver operate at, roughly, the same speed.

For asynchronous transmission the start and stop bits are added in addition to the seven ASCII character bits and the parity. Thus a total of 10
bits are required to transmit a single character. With two stop bits, a total

of 11 bits are required. When sending 10 characters every second and if 11 bits are used for each character, then the transmission rate is 110 bits per second (bps). Table 8.3 lists how the bit rate relates to the characters sent per second (assuming 10 transmitted bits per character). The bit rate is measured in bits per second (bps).

	Bits
ASCII character	7
Start bit	1
Stop bit	2
Total	10

Table 8.3 Bits per second related to characters sent per second

Speed(bps)	Characters / second
300	30
1200	120
2400	240

In addition to the bit rate, another term is used to describe the transmission speed is the baud rate. The bit rate refers to the actual rate at which bits are transmitted, whereas, the baud rate relates to the rate at which signalling elements, used to represent bits, are transmitted. Since one signalling element encodes one bit, the two rates are then identical. Only in modems does the bit rate differ from the baud rate.

8.5.3 Bit stream timings

Asynchronous communications is a stop-start mode of communication and both the transmitter and receiver must be set up with the same bit timings. A start bit identifies the start of transmission and is a low logic level. Next the least significant bit is sent followed by the rest of the bits in the character. After this the parity bit is sent followed by the stop bit(s). The actual timing of each bit is related to the baud rate and can be found using the following formula:

$$\text{Time period of each bit} = \frac{1}{\text{baud rate}} \text{ s}$$

For example, if the baud rate is 9600 baud (or bps) then the time period for each bit sent is 1/9600 s or 104 μs. Table 8.4 shows some bit timings as related to baud rate. An example of the voltage levels and timings for the ASCII character 'V' is given in Figure 8.11.

Table 8.4   Bit timings related to baud rate

Baud rate	Time for each bit (μs)
1200	833
2400	417
9600	104
19200	52

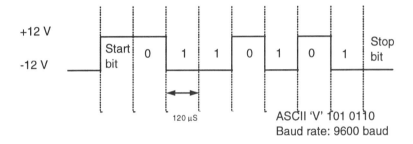

Figure 8.11   ASCII 'V' at RS-232 voltage levels

8.6 STANDARDS

8.6.1 Standards organizations

The main standards organizations for data communications are the ITU (International Telecommunications Union), the EIA (Electronic Industry Association) and the ISO (International Standards Organisation). The ITU standards related to serial communications, are defined in the V-series specifications and EIA standards as the RS-series.

8.6.2 EIA standard RS-232-C

The RS-232-C standard defines the interfacing of a DTE to a DCE over a distance of up to 50 feet and at a maximum data rate of 20 kbps.

8.6.3 EIA standard RS-449, RS-422A, RS-423A

RS-232 interfaces computer/data terminal equipment separated by a distance up to 50 feet. The EIA has since generated three standards which improve the specification of the interconnection giving higher data rates and longer maximum interconnection lengths. RS-422 and RS-423 define electrical characteristics while RS-499 defines the basic interface standards and refers to the RS-422/3 standards. These standards are:

- RS-422A (electrical characteristics of balanced load voltage digital interface circuits);
- RS-423A (electrical characteristics of unbalanced voltage digital interface circuits);
- RS-449 (general purpose 37-position and 9-position interface for DTE and DCE employing serial binary data interchange).

8.6.4 EIA standard RS-485

RS-485 is an upgraded version of RS-422 and extends the number of peripherals that can be interfaced. It allows for bi-directional multi-point party line communications. This can be used in networking applications. RS-422 and RS-232 facilitate simplex communication, whereas RS-485 allows for multiple receivers on a single line facilitating half-duplex communications. The maximum data rate is unlimited and is set by the rise time of the pulses, but it is usually limited to 10 Mbps. A network using the RS-485 standard can have up to 32 transmitters/receivers with a maximum cable length of 1.2 km, as shown in Figure 8.12.

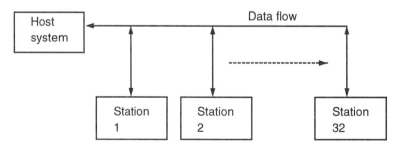

Figure 8.12   RS-485 connecting to multiple nodes.

8.7 LINE DRIVERS

Transmission lines have effects on digital pulses in the following ways:

- Attenuation – The transmission line contains series resistance which causes a reduction in the pulse amplitude.
- Pulse distortion – The transmission line insulation produces a shunt capacitance on the signal path and a series resistance and inductance of the conductors. This causes the transmission line to distort the shape of the pulse. The two main effects are the block of high frequencies in the pulse and phase distortion.
- Noise – Noise is any unwanted electrical signals added to a signal. A digital system is less prone to noise as it has only two levels and it takes a relatively large change in voltage to cause an error.

Table 8.5 shows the electrical characteristics of the different serial communication standards. The two main standards agencies are the EIA and the ITU.

Balanced lines use two lines for each signal line, whereas, unbalanced lines use one wire for each signal and a common return circuit (see Figure 8.13). RS-422 is a balanced interface and uses two conductors to carry the signal (see Figure 8.14). The electrical current in each of the conductors is 180° out-of-phase with each other. Balanced lines are generally less prone to noise as any noise induced into the conductors will be of equal magnitude. At the receiver the noise will tend to cancel out.

The voltage levels for RS-232 range from ±3 to ±25 V, whereas, for RS422/RS423 the voltage ranges are ±0.2 to ±6 V. For very high bit rates the cable is normally terminated with the characteristic impedance of the line, for example a 50Ω cable is terminated with a 50Ω termination.

RS-422 interface circuits can have up to 10 receivers. They have no ground connection and are thus useful in isolating two nodes. For two-way communications four connections are required, the TX+ and TX– on one node connects to the RX+ and RX– on the other.

Nodes may have a direct RS-422 connection or can be fitted with a special interface connector to convert from RS-232 to RS-422 (although the maximum data rate is likely to be limited to the maximum RS-232 rate).

It should also be noted that the maximum connection distance relates to the maximum data rate. If a lower data rate is used then the maximum distance can be increased. For example, in some situations with a good quality cable and in a low noise environment, it is possible to have cable run of 1 km using RS-232 at 1200 bps.

Table 8.5   Main serial standards

EIA	RS-232-C	RS-423-A	RS-422-A	RS-485
ITU	V.28	V.10/X.26	V.11/X.27	
Data rate	20 kbps	300 kbps	10 Mbps	10 Mbps
Max. Distance	15 m	1200 m	1200 m	1200 m
Type	Unbalanced	Unbalanced Differential	Balanced Differential	Balanced Differential
No. of drivers and receivers	1 driver 1 receiver	1 driver 10 receivers	1 driver 10 receivers	32 drivers 32 receivers
Driver voltages	±15 V	± 6 V	±5 V	±5 V
No. of conductors per signal	1	2	2	2

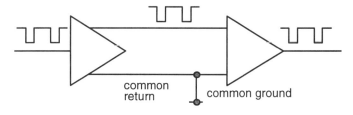

Figure 8.13   Unbalanced digital interface circuit (RS-423).

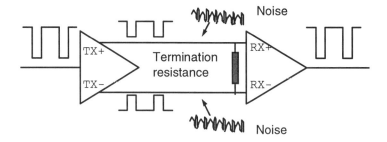

Figure 8.14   Balanced digital interface circuit (RS-422).

8.8 COMMUNICATIONS BETWEEN TWO NODES

RS-232 is intended to be a standard but not all manufacturers abide by it. Some implement the full specification while others implement just a partial specification. This is mainly because not every device requires the full functionality of RS-232, for example a modem requires many more control lines than a serial mouse.

The rate at which data is transmitted and the speed at which the transmitter and receiver can transmit/receive the data dictates whether data handshaking is required.

8.8.1 Handshaking

In the transmission of data there can be no either handshaking, hardware handshaking or software handshaking. If no handshaking is used then the receiver must be able to read the received characters before the transmitter sends another. The receiver may buffer the received character and store it in a special memory location before it is read. This memory location is named the receiver buffer. Typically, it may only hold a single character. If it is not emptied before another character is received then any character previously in the buffer will be overwritten. An example of this is illustrated in Figure 8.15. In this case the receiver has read the first two characters successfully from the receiver buffer, but it did not read the third character as the fourth transmitted character has overwritten it in the receiver buffer. If this condition occurs then some form of handshaking must be used to stop the transmitter sending characters before the receiver has had time to service the received characters.

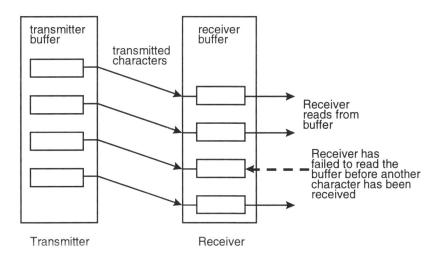

Figure 8.15 Transmission and reception of characters.

Hardware handshaking involves the transmitter asking the receiver if it is ready to receive data. If the receiver buffer is empty it will inform the transmitter that it is ready to receive data. Once the data is transmitted and loaded into the receiver buffer the transmitter is informed not to transmit any more characters until the character in the receiver buffer has been read. The main hardware handshaking lines used for this purpose are:

- CTS – Clear To Send;
- RTS – Ready To Send;
- DTR – Data Terminal Ready;
- DSR – Data Set Ready.

Software handshaking involves sending special control characters. These include the DC1-DC4 control characters.

8.8.2 RS-232 pocket tester and break-out boxes

An RS-232 pocket tester is useful in determining which handshaking lines are active and which are inactive. They are normally self-powered and have 25-pins which are wired straight through so that it can be connected in-line from one device to another. The lines connect monitored are pins 2, 3, 4, 5, 6, 8 and 20 which monitor the lines TD, RD, RTS, CTS, DSR, CD and DTR. They are dual colour LEDs and a Green light indicates a low voltage input (a 1, or a MARK) and red indicates a high voltage input (a 0 or a SPACE).

A break-out box allows any of the signal lines to be connected to either side of the connection. An RS-232 pocket tester and micro break-out box are illustrated in Figure 8.16.

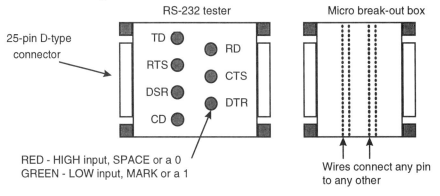

Figure 8.16   RS-232 tester and micro break-out box.

8.8.3 RS-232 set-up

Figure 8.17 shows a sample set-up taken from the Terminal program available with Microsoft Windows. The selectable baud rates are 110, 300, 600, 1200, 2400, 4800, 9600 and 19 200 baud. Notice that the flow control can either be set to software handshaking (Xon/Xoff), hardware handshaking or none.

The parity bit can either be set to none, odd, even, mark or space. A mark in the parity option sets the parity bit to a '1' and a space sets it to a '0'.

In this case COM1: is set at 1200 baud, 8 data bits, no parity, 1 stop bit and no parity checking. Notice that with this package a parity error is ignored unless the Parity Check box is activated.

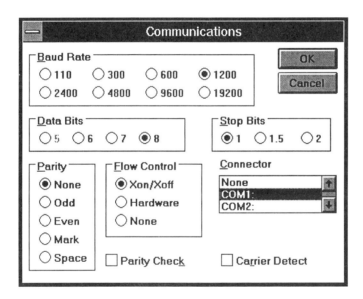

Figure 8.17   Sample communications set-up screen from Microsoft Terminal.

8.8.4 Simple no-handshaking communications

In this form of communication it is assumed that the receiver can read the received data from the receive buffer before another character is received. Data is sent from a TD pin connection of the transmitter and is received in the RD pin connection at the receiver. When a DTE (such as a computer) connects to another DTE, then the transmit line (TD) on one is connected to the receive (RD) of the other and vice versa. Figure 8.18 shows the connections between the nodes.

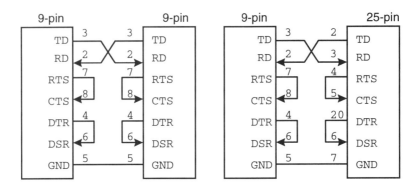

Figure 8.18   RS-232 connections with no hardware handshaking.

8.8.5 Software handshaking

There are two ASCII characters that start and stop communications. These are X-ON (^S , Cntrl-S or ASCII 11) and X-OFF (^Q, Cntrl-Q or ASCII 13). When the transmitter receives an X-OFF character it ceases communications until an X-ON character is sent. This type of handshaking is normally used when the transmitter and receiver can process data relatively quickly. Normally, the receiver will also have a large buffer for the incoming characters. When this buffer is full it transmits an X-OFF. After it has read from the buffer the X-ON is transmitted, see Figure 8.19.

8.8.6 Hardware handshaking

Hardware handshaking stops characters in the receiver buffer from being overwritten. The control lines used are all active HIGH. Figure 8.20 shows how the nodes communicate. When a node wishes to transmit data it asserts the RTS line active (that is, HIGH). It then monitors the CTS line until it goes active (that is, HIGH). If the CTS line at the transmitter stays inactive then the receiver is busy and cannot receive data, at the present. When the receiver reads from its buffer the RTS line will automatically goes active indicating to the transmitter that it is now ready to receive a character.

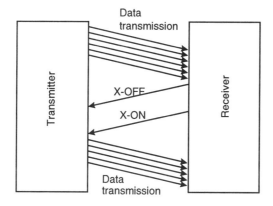

Figure 8.19   Software handshaking using X-ON and X-OFF.

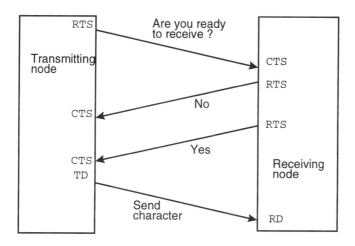

Figure 8.20   Handshaking lines used in transmitting data.

Receiving data is similar to the transmission of data, but, the lines DSR and DTR are used instead of RTS and CTS. When the DCE wishes to transmit to the DTE the DSR input to the receiver will become active. If the receiver cannot receive the character it will set the DTR line inactive. When it is clear to receive it sets the DTR line active and the remote node then transmits the character. The DTR line will be set inactive until the character has been processed.

8.8.7 Two-way communications with handshaking

For full handshaking of the data between two nodes the RTS and CTS lines are crossed over (as are the DTR and DSR lines). This allows for full remote node feedback (see Figure 8.21).

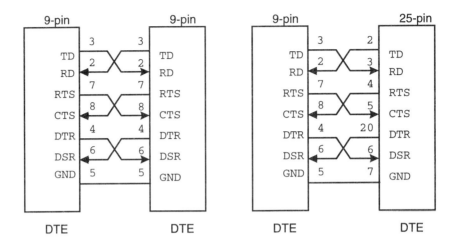

Figure 8.21   RS-232 communications with handshaking.

8.8.8 DTE-DCE connections

A further problem occurs in connecting two nodes. A DTE-DTE connection requires crossovers on their signal lines, whereas DTE-DCE connections require straight-through lines. An example connection for a computer to modem connection is shown in Figure 8.22.

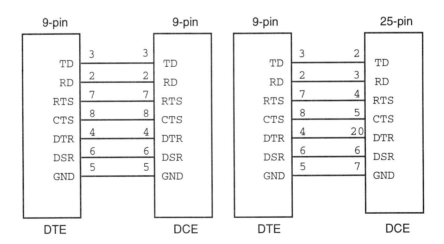

Figure 8.22   DTE to DCE connections

8.9 PRACTICAL RS-232 PROGRAM

Practical RS-232 programs for a PC-based system are given in Appendix A.

8.10 ASYNCHRONOUS PROTOCOLS

Asynchronous protocols are character-oriented and are typically used with asynchronous communications to provide flow and error control. Unique characters are sent between nodes to acknowledge the reception (or not) of data, when an error occurs and so on. Table 8.6 outlines some specially reserved ASCII characters which provide flow information.

Once communication has been started a typical message transfer is shown in Figure 8.23. In this case when the transmitter is ready to transmit it polls the receiver with an ENQ character. If the receiver is ready it will reply with an acknowledgement character (ACK) else it sends NAK. When the transmitter is then ready to send data it sends a start of transmission character (STX) then it sends data. After data is sent the receiver can reply with an acknowledgement (ACK) if it received the data or a NAK if it did not. If a NAK is received then the transmitter may send the data again. At the end of the transmission the transmitter sends an end of transmission character (EOT).

Table 8.6   Flow control ASCII characters

Symbol	Function	ASCII code
EOT	End of Transmission. End of text or messages	04h
STX	Start of Transmission indicates start of actual text	02h
ACK	Acknowledge. Positive acknowledgement of receipt or responding to a poll	06h
ENQ	Enquiry. Requesting for a response, i.e. 'have you anything to send'	05h
NAK	Negative Acknowledge. This signifies an error in a received message	15h
SYN	Synchronous Idle. This character provides synchronization of bits.	16h

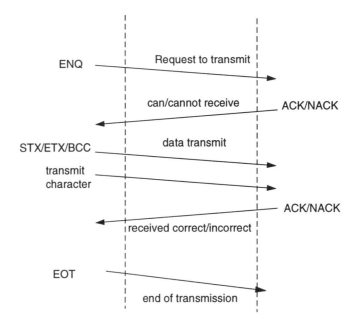

Figure 8.23   Typical asynchronous communications

8.11 EXERCISE

8.1   Which organization defined the original RS-232C standard:

        A   ANSI
        B   IEEE
        C   CCITT
        D   EIA

8.2   Which organization defined the V.24 standard:

        A   ANSI
        B   IEEE
        C   CCITT
        D   EIA

8.3   The main advantage of serial communications over parallel communications is that:

A   it is faster
B   it has less errors
C   it requires less transmission wires
D   it is easier to send the data

8.4   Which of the following best describes simplex communications:

A   one-way communication
B   two-way communication, one node at a time
C   two-way communications, two nodes at the same time
D   none of the above

8.5   Which of the following best describes full-duplex communications:

A   one-way communication
B   two-way communication, one node at a time
C   two-way communications, two nodes at the same time
D   none of the above

8.6   Which of the following devices is defined as a DTE

A   a modem
B   an RS-232 cable
C   a computer
D   an RS-232 connector

8.7   Nominally, what voltage is a logic 0 on an RS-232 line:

A   0 V
B   +12 V
C   −12 V
D   +5 V

8.8   Nominally, what voltage is a logic 1 on an RS-232 line:

A   0 V
B   +12 V
C   −12 V
D   +5 V

8.9   A line-break would be detected at the receiver as which voltage level:

A   0 V
B   +12 V
C   −12 V
D   +5 V

8.10   The inactive state on an RS-232 line is:

A   0 V
B   +12 V
C   −12 V
D   +5 V

8.12 TUTORIAL

8.11   Decode the following RS-232 transmitted message:

First bit sent
⇓
11111111101100111111111101000011111101011011

```
11100001101011100000010111111111101001111111
11111001001111110010011111111110101001111101
011001111111
```

Assume ASCII coding, one stop bit and ignore the parity bit.

Refer to Appendix A for practical RS-232 tutorials

9

Modems

9.1 INTRODUCTION

The word 'modem' is a contraction of MOdulator/DEModulator. A modem connects digital equipment to a speech bandwidth-limited communications channel. Typically, they are used on telephone lines which have a bandwidth of between 400 Hz and 3.4 kHz. If digital pulses were applied directly these lines they would end up severely distorted.

If the modem connects to the public telephone line it should normally be able to do the following:

- automatically dial another modem using either touch-tone or pulse dialling;
- automatically answer calls and make a connection with another modem;
- disconnect a phone connection when data transfer has completed or if an error occurs;
- convert bits into a form suitable for the line (modulator);
- convert received signals back into bits (demodulator);
- software or hardware handshaking of data or no handshaking.

Figure 9.1 shows how two computers connect to each other using RS-232 converters and modems. The RS-232 converter is normally an integral part of the computer, while the modem can either be external or internal to the computer. If it is externally connected then it is normally connected by a cable with a 25-pin male D-type connector on either end.

There are two types of circuits available from the public telephone network:

- direct distance dialling (DDD).
- private line (or leased line).

The DDD is a dial-up network where the link is established in the same manner as normal voice calls with a standard telephone or some kind of an automatic dial/answer machine. They can use either touch-tones or pulses to make the connection. With private line circuits, the subscriber has a permanent dedicated communication link.

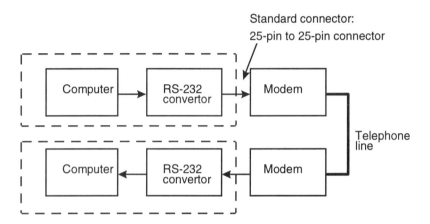

Figure 9.1   Computer communications using a modem

Modems are either synchronous or asynchronous. A synchronous modem recovers the clock at the receiver. There is no need for start and stop bits in a synchronous modem. Asynchronous modems are by far the most popular types. A measure of the speed of the modem is the baud rate or bps (bits per second).

9.2 DIGITAL MODULATION

Digital modulation changes the characteristic of a carrier according to binary information. With a sine-wave carrier thc amplitude, frequency or phase can be varied. The three basic types of digital modulation are:

- amplitude-shift keying (ASK);
- frequency-shift keying (FSK);
- phase-shift keying (PSK).

The three basic types are shown in Figure 9.2.

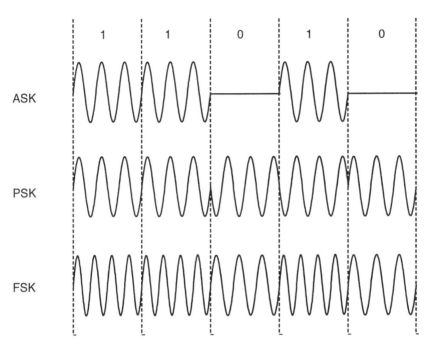

Figure 9.2 Waveforms for ASK, PSK and FSK

9.2.1 Frequency shift keying (FSK)

FSK, in the most basic case, represents a 1 (a mark) by one frequency (a mark) and a 0 (a space) by another. These frequencies lie within the bandwidth of the transmission channel.

On a V.21, 300 bps, full-duplex modem the originator modem uses the frequency 980 Hz to represent a mark and 1180 Hz a space. The answering modem transmits with 1650 Hz for a mark and 1850 Hz for a space. This is illustrated in Figure 9.3. The four frequencies allow the caller originator and the answering modem to communicate at the same time, that is full-duplex communication.

FSK modems are inefficient in their use of bandwidth, with the result that the maximum data rate over normal telephone lines is 1800 bps. Typically, for rates over 1200 bps, other modulation schemes are used.

9.2.2 Phase shift keying (PSK)

In coherent PSK a carrier gets no phase shift for a 0 and a 180° phase shift for a 1, as given next:

$$0 \Rightarrow 0°$$
$$1 \Rightarrow 180°$$

Its main advantage over FSK is that since it uses a single frequency it uses much less bandwidth. It is thus less affected by noise. It has the advantage over ASK in that the information is not contained in the amplitude of the carrier, thus again it is less affected by noise.

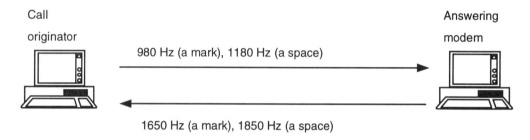

Call originator

Answering modem

980 Hz (a mark), 1180 Hz (a space)

1650 Hz (a mark), 1850 Hz (a space)

Figure 9.3 Waveforms for ASK, PSK and FSK

9.2.3 M-ary modulation

With M-ary modulation a change in amplitude, phase of frequency represents one of M possible signals. It is possible to have M-ary FSK, M-ary PSK and M-ary ASK modulation schemes. This is where the baud rate differs from the bit rate. The bit rate is the true measure of the rate of the line, whereas the baud rate only indicates the signalling element rate, which might be, typically, a half or a quarter of the bit rate.

For four-phase differential phase shift keying (DPSK) the bits are grouped into 2 bits and each group is assigned a certain phase shift. For 2 bits there are four combinations: a 00 is coded as 0°, 01 coded as 90°, and so on.

$$00 \Rightarrow 0° \quad 01 \Rightarrow 90°$$
$$11 \Rightarrow 180° \quad 10 \Rightarrow 270°$$

It is also possible to change a mixture of amplitude, phase or frequency. M-ary amplitude-phase keying (APK) varies both the amplitude and phase of a carrier to represent M possible bit patterns.

M-ary quadrature amplitude modulation (QAM) changes the amplitude and phase of the carrier. 16-QAM uses four amplitudes and four phase shifts, it can thus be used to code 4 bits at a time. In this case the Baud rate will be a quarter of the bit rate.

Typical technologies for modems are given next:

FSK   – used up to 1200 bps
Four-phase DPSK – used at 2400 bps
Eight-phase DPSK – used at 4800 bps
16-QAM   – used at 9600 bps

9.3 MODEM STANDARDS

The CCITT (now known as the ITU) have defined standards which relate to RS-232 and modem communications. Each uses a V. number to define their type, as given in Table 9.1.

Most currently available modems use V.21, V.22, V.22bis, V.23 and V.32 line speeds and are Hayes compatible. Hayes was the company that pioneered modems and defined a standard method of programming the mode of the modem with the AT command language.

Table 9.1   CCITT V. series standards

CCITT	Description
V.21	Full-duplex modem transmission at 300 bps.
V.22	Half-duplex modem transmission at 600 bps and 1200 bps.
V.22bis	Full-duplex modem transmission at 1200 bps and 2400 bps.
V.23	Full-duplex modem transmission at 1200 bps and receive at 75 bps.
V.24	The CCITT standard for the RS-232 interface.
V.32	Full-duplex modem transmission at 4800 and 9600 bps.
V.25bis	Modem command language.
V.32bis	Full-duplex modem transmission at 7200, 12000 and 14400 bps.
V.42	Error control protocol.

9.4 MODEM COMMANDS

A computer gets the attention of the modem by sending an 'AT' command. For example, 'ATDT' is the dial command. Initially, a modem is in command mode and accepts commands from the computer. These commands are sent at either   300 bps   or   1200 bps   (the modem automatically detects which of the speeds is being used).

Most commands are sent with the AT prefix. Each command is followed by a carriage return character (ASCII character 13 decimal); a command without a carriage return character is ignored. More than one command can be placed on a single line and, if necessary, spaces can be entered to improve readability. Commands can be sent either in upper or lower case. Example AT commands are listed in Table 9.2.

Table 9.2 Example AT modem commands

Command	Description
ATDT12345	Automatically phone number 12345 using touch-tone dialling.
ATPT12345	Automatically phone number 12345 using pulse dialling.
AT S0=2	Automatically answer a call. The S0 register contains the number of rings the modem uses before it answers the call. In this case there will be two rings before it is answered. If S0 is zero then the modem will not answer a call.
ATH	Hang-up telephone line connection.
+++	Disconnect line and return to on-line command mode.

The modem can enter into one of two states: the normal state and the command state. In the normal state the modem transmits and/or receives characters from the computer. In the command state, characters sent to the modem are interpreted as commands. Once a command is interpreted the modem goes into the normal mode. Any characters sent to the modem are then sent along the line. To interrupt the modem so that it goes back into command mode, three consecutive '+' characters are sent, that is, '+++'.

The modem contains various status registers called the S-registers which store modem settings. For example the S0 register stores the number of rings that the modem receives before it answers a call. Table 9.3 lists some of these registers.

Most modems have an area of non-volatile random access memory (NVRAM) which can be programmed electronically. This allows the modem to automatically power up in a pre-programmed mode.

After the modem has received an AT command it responds with a return code. Example return codes are given in Table 9.4. For example if a modem calls another which is busy then the return code is 7. A modem dialling another modem returns the codes for OK (when the ATDT command is received), CONNECT (when it connects to the remote modem) and CONNECT 1200 (when it detects the speed of the remote modem). Note that the return code from the modem can be suppressed by sending the AT command 'ATQ1'. The AT code for it to return the code is 'ATQ0', normally this is the default condition

Table 9.3 Modem registers

Register	Function	Range [typical default]
S0	Rings to Auto-answer	0–255 rings [0 rings]
S1	Ring counter	0–255 rings [0 rings]
S2	Escape character	[43]
S3	Carriage return character	[13]
S6	Wait time for dial tone	2–255 s [2 s]
S7	Wait time for carrier	1–255 s [50 s]
S8	Pause time for automatic dialing	0–255 [2 s]
S32	XON character	0–255 [11h]
S33	XOFF character	0–255 [13h]

Table 9.4 Example return codes

Message	Digit	Description
OK	0	Command executed without errors
CONNECT	1	A connection has been made
RING	2	A incoming call has been detected
NO CARRIER	3	No carrier detected
ERROR	4	Invalid command
CONNECT 1200	5	Connected to a 1200 bps modem
NO DIALTONE	6	Dial tone not detected
BUSY	7	Remote line is busy
NO ANSWER	8	No answer from remote line
CONNECT 600	9	Connected to a 600 bps modem
CONNECT 2400	10	Connected to a 2400 bps modem
CONNECT 4800	11	Connected to a 4800 bps modem

9.5 MODEM CONNECTIONS

RS-232C is a standard that defines the signal functions and their electrical characteristics for connecting a DTE (data terminal equipment, such as, a computer) and a DCE (data circuit termination equipment, such as, a modem). CCITT has adopted the RS-232C standard as V.24 for the signal functions and V.28 as the electrical characteristics. The most common connector on the cable for this type of communication is a 25-pin D-type male connector.

The computer sends data to the modem via the TX data line and receives it from the RX data line. A ground line is also required, see Figure 9.4.

The modem can either use no handshaking, flow control handshaking or hardware handshaking. With hardware handshaking the sending computer sets the Read To Send (RTS) signal on the modem. It then replies, when ready, with a Clear To Send (CTS). Data Set Ready (DSR) from the modem and Data Terminal Ready (DTR), from the computer, work in the same way but for received data.

With no handshaking or flow control handshaking, the RTS and DTR signals are normally set active by the computer to inform the modem that it is ready to transmit and/or receive data continuously.

The Ring Indicator (RI) line is active when the modem detects an incoming call. The Data Carrier Detect (DCD) line is active when a connection has been established. DTR is used to control most modems for auto-answer dialling and hang-up circuits; it must be active to auto-answer and if it is inactive then the modem hangs up.

Figure 9.5 shows a sample window from the Microsoft Windows Terminal program. It shows the Modem commands window. In this case, it can be seen that when the modem dials a number the prefix to the number dialled is 'ATDT'. The hang-up command sequence is '+++ ATH'. A sample dialling window is shown in Figure 9.6. In this case the number dialled is 9,4567890. A ',' character represent a delay. The actual delay is determined by the value in the S8 register (see Table 9.3). Typically, this value is about two seconds.

On many private switched telephone exchanges a 9 must prefix the number if an outside line is required. A delay is normally required after the 9 prefix before dialling the actual number. To modify the delay to five seconds, dial the number 9 0112432 and to wait 30 seconds for the carrier, then the following command line can be used:

```
ATDT 9,0112432 S8=5 S7=30
```

It can be seen in Figure 9.5 that a prefix and a suffix is sent to the modem.

This is to ensure that there is a time delay between the transmission prefix and the suffix string. For example when the modem is to hang up the connection, the '+++' is sent followed by a delay then the 'ATH'.

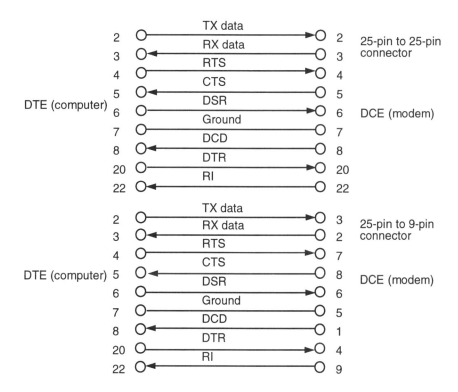

Figure 9.4 Modem line connections

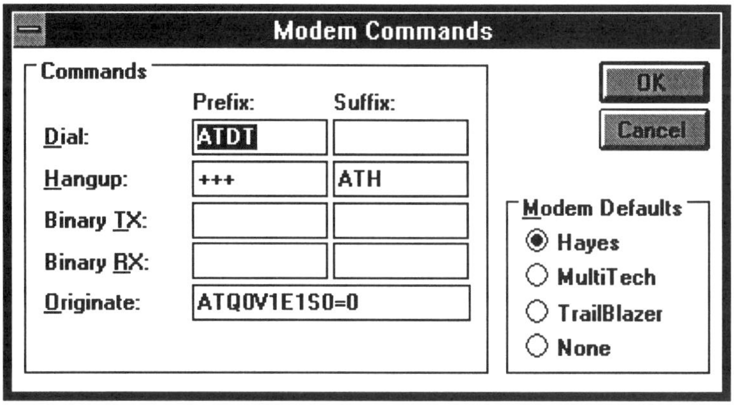

Figure 9.5 Modem commands

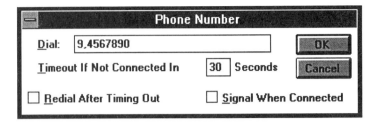

Figure 9.6   Dialling a remote modem

In Figure 9.5 there is an option for the Originator. This is the string that is sent initially to the modem to set it up. In this case the string is 'ATQ0V1E1S0=0'. The Q0 part informs the modem to return a send status code. The V1 part informs the modem that the return code message is to be displayed rather than just the value of the return code; for example, it displays CONNECT 1200 rather than the code 5 (V0 displays the status code). The E1 part enables the command message echo (E0 disables it).

9.6 MODEM INDICATORS

Most external modems have status indicators to inform the user of the current status of a connection. Typically, the indicator lights are:

- AA – is ON when the modem is ready to receive calls automatically. It flashes when a call is incoming. If it is OFF then it will not receive incoming calls. Note that if the S0 register is loaded with any other value than 0 then the modem goes into auto-answer mode. The value stored in the S0 register determines the number of rings before the modem answers.
- CD – is ON when the modem detects the remote modem's carrier, else it is OFF.
- OH – is ON when the modem is on-hook, else it is OFF.
- RD – flashes when the modem is receiving data or is getting a command from the computer.
- SD – flashes when the modem is sending data.
- TR – shows that the DTR line is active (that is, the computer is ready to transmit or receive data).
- MR – shows that the modem is powered on.

9.7 TUTORIAL

9.1   Find a PC with Microsoft Windows and run the Terminal program (normally found in the Accessories group). Determine the following:

(a)   the default RS-232 settings, such as baud rate, the parity, flow control, and so on (select Communications... from the Settings menu);

(b)   the hang-up command for the modem (select Modem Command... from the Settings menu);

(c)   the dial-up command for the modem (select Modem Command... from the Settings menu).

9.2   Explain how a DTE (a computer) gets the attention of the modem. Also, explain how the DTE gets the modem to go into the command mode once it is in normal mode.

9.3   Which modem indicators would be ON when a modem has made a connection and is receiving data? Which indicators will be flashing?

9.4   Which modem indicators would be ON when a modem has made a connection and is sending data? Which indicators will be flashing?

9.5   Investigate the complete set of AT commands by referring to a modem manual or reference book.

9.6   Investigate the complete set of S-registers by referring to a modem manual or reference book.

9.7   Determine the location of modems on a network or in a works building. If possible, determine the type of data being transferred and its speed.

9.8   If possible, connect a modem to a computer and dial a remote modem.

9.9   If possible connect two modems together and, using a program such as Terminal, transfer text from one computer to the another.

10

Pulse coded modulation (PCM)

10.1 INTRODUCTION

Pulse coded modulation involves analogue data being transmitted as digital signals. The digital form is then transmitted over the transmission media. At the receiver the digital code converts back into an analogue form. Digital signals are generally less affected by noise and thus provide a more reliable method of transmitting data. Figure 10.1 shows an example digital signal with noise added to it. The comparator outputs a HIGH level (a '1') if the signal voltage is greater than the threshold voltage, or it will output a LOW. If the noise voltage is less than half the threshold voltage then the noise will not affect the recovered signal. Even if the noise is greater than this threshold there are techniques which can reduce its effect. One method is to add extra bits to the data to either detect errors or even to correct the bits in error.

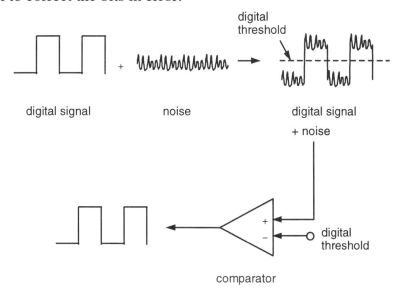

Figure 10.1   Recovery of a digital signal with noise added to it

The accuracy of the PCM depends on the number of bits used for each analogue sample. This gives the PCM a dependable response over an equivalent analogue system because an analogue system's accuracy depends on component tolerance, producing a differing response for different systems.

It is difficult, if not impossible, to recover the original analogue signal after the addition of noise, especially if the noise is random. Most methods of reducing noise involve some form of filtering or smoothing of the signal.

10.1.1 Sampling theory

As a signal may be continually changing a sample of it must be taken at given time intervals. The rate of sampling depends on its rate of change. For example, the temperature of the sea will not vary much over a short time but a video image of a sports match will. To encode a signal digitally it is normally sampled at fixed time intervals. Sufficient information is then extracted to allow the signal to be processed or reconstructed.

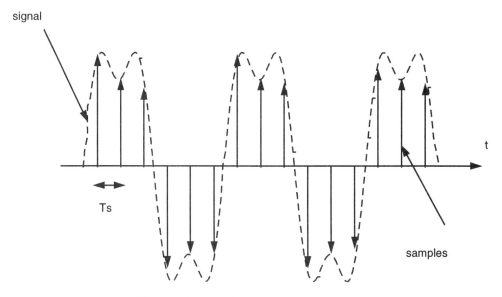

Figure 10.2 The sampling process

If a signal is to be reconstructed as the original signal it must be sampled at a rate defined by the Nyquist Criterion. This states that the sampling rate must be twice the highest frequency of the signal. For telephone speech channels of 4 kHz, the signal must be sampled at least 8 000 times per second (8 kHz), that is, once every 125 μs. Figure 10.2 shows a signal sampled every T_S seconds.

10.1.2 Quantization

Before an analogue signal is coded into a digital form it is first quantized into discrete steps of amplitude. Once quantized, the instantaneous values of the continuous signal can never be exactly restored. This leads to random error called quantization error. Figure 10.3 shows a sample quantization process. Notice that the 11th and the 12th sample are both coded as binary 0001, although, they differ in amplitude.

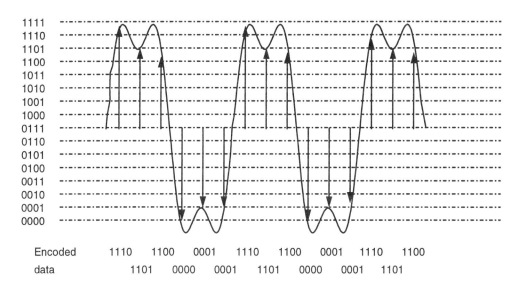

Figure 10.3   Sampling and quantization

10.1.3 Coding of samples

The more quantization levels the smaller the level gap and the more accurate the final coded value. Figure 10.4 shows examples of uni-polar and bi-polar three-bit coding. In uni-polar coding the signal is always positive, whereas, in bi-polar conversion the signal has positive and negative values. The full range of the input signal fits between 0 and FS (Full Scale) for uni-polar and –FS/2 to +FS/2 for bi-polar. Typically this the range will be from 0 V to +5, +10 or +15 V for uni-polar and ±5 or ±15 V for bi-polar. The number of quantization level depends on the number of bits used. In this case with 3 bits there will be eight different coded values from 000 to 111.

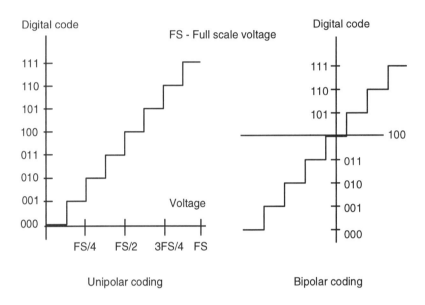

Figure 10.4   3-bit uni-polar and bipolar coding

The most significant bit represents the full-scale voltage divided by 2, the next significant bit represents the full scale voltage divided by 4 and so on to the least significant bit which represents the full-scale voltage divided by 2^N. The number of quantization levels is 2^N. Table 10.1 shows how the accuracy varies with the number of the quantization levels.

Table 10.1   Number of quantization levels as a function of bits

Bits (n)	Quantization levels	Accuracy (%)
1	2	50
2	4	25
3	8	12.5
4	16	6.25
8	256	0.2
12	4096	0.012
14	16384	0.003
16	65536	0.000 76

10.1.4 Quantization error

The maximum error between the original level and the quantized level occurs when the original level falls exactly half way between two quantized levels. This error will be a half of the smallest increment or

$$\text{Max error} = \pm\frac{1}{2}\frac{\text{Full scale}}{2^N}.$$

10.2 PCM PARAMETERS

The main parameters in determining the quality of a PCM system are the dynamic range and signal-to-noise ratio.

10.2.1 Dynamic range (DR)

The dynamic range is the ratio of the largest possible signal magnitude to the smallest possible magnitude. If the input signal uses the full range of the ADC then the maximum signal will be the full-scale voltage. The smallest signal which can be reproduced is one which toggles between one quantization level and the level above, or below. This signal amplitude, for an n-bit ADC, is the full-scale voltage divided by the number of quantization levels (that is, 2^n). Thus, for a linear quantized signal:

$$\text{Dynamic Range} = \frac{V_{max}}{V_{min}}$$

$$\text{Number of levels} = 2^n - 1$$

$$\text{Dynamic Range} = 20\log\frac{V_{max}}{V_{max}/{2^n - 1}} \quad \text{dB}$$

$$= 20\log(2^n - 1) \quad \text{dB}$$

if $2^n \gg 1$ then

$$\text{Dynamic Range} = 20\log 2^n \quad \text{dB}$$
$$= 20n\log 2 \quad \text{dB}$$
$$= 6.02n \quad \text{dB}$$

10.2.2 Signal-to-noise ratio (SNR)

It can be shown that the SNR for a linearly quantized digital system is

$$SNR = 1.76 + 6.02n \ dB$$

For example, a 16-bit system has a SNR of 98.08 dB and a DR of 96.32 dB.

10.3 SPEECH COMPRESSION

Subjective and system tests have found that 12-bit coding is required to code speech signals, which gives 4096 quantization levels. If linear quantization is applied then the quantization step is the same for quiet levels as for loud levels. Any quantization noise in the signal will be more noticeable at quiet levels than at loud levels. When the signal is loud, the signal itself swamps the quantization noise, as illustrated in Figure 10.5. Thus an improved coding mechanism is to use small quantization steps at low input levels and a higher one at high levels. This is achieved using non-linear compression.

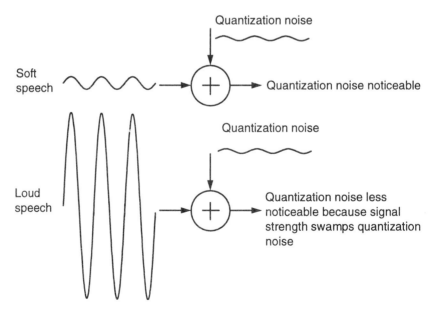

Figure 10.5   Quantization noise is more noticeable with low signal levels

The two most popular types of compression are A-Law (in European systems) and μ-Law (in the USA). These laws are similar and compress the 12-bit quantized speech code into an 8-bit compressed code. An example compression curve is shown in Figure 10.6.

As an approximation the two laws are split into 16 line segments. Starting from the origin, each segment, outwards, is half the slope of the previous.

Using an 8-bit compressed code at a sample rate of 8000 samples per second gives a bit-rate of 64 kbps. ISDN uses this bit rate to transmit digitized speech. Figure 10.7 shows a basic transmission system.

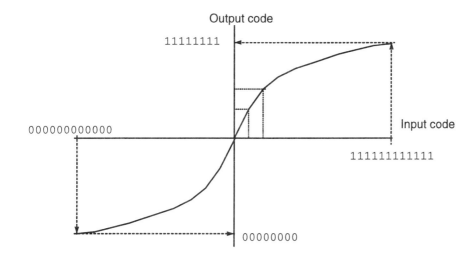

Figure 10.6 12-bit to 8-bit non-linear compression

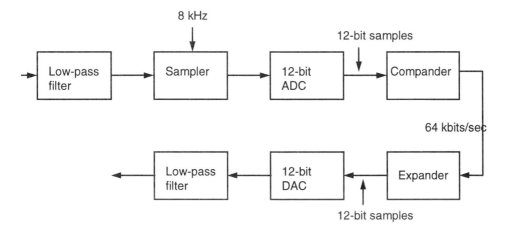

Figure 10.7 Typical PCM speech system

10.4 DELTA MODULATION PCM

Delta modulation uses a single-bit PCM code to represent the analogue signal. With delta modulation a '1' is transmitted if the analogue input is higher than the previous sample or a '0' if it is lower. It must obviously work at a higher rate than the Nyquist frequency but because it only uses 1 bit it normally uses a lower output bit-rate. A delta modulation transmitter is shown in Figure 10.8.

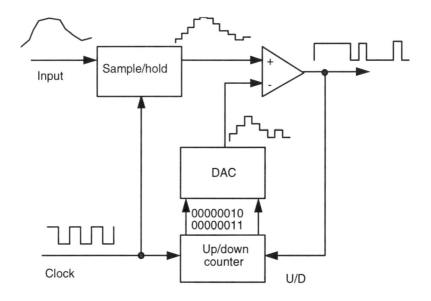

Figure 10.8 Delta modulation

Initially the counter is set to zero. A sample is taken and if it is greater than the analogue value on the DAC output then the counter is increment by 1, or it is decremented. This continues at a time interval given by the clock. Each time the present sample is greater than the previous then a '1' is transmitted, or a '0' is transmitted. Figure 10.9 shows an example signal. The sampling frequency is chosen so that the tracking DAC can follow the input signal. This results in a higher sampling frequency, but because it only transmits one bit at a time then the output bit rate is normally reduced.

Figure 10.10 shows that the receiver is almost identical to the transmitter except that it has no comparators.

Two problems with delta modulation are granular noise and slope overload. Slope overload occurs when the signal changes too fast for the modulator to keep up with, see Figure 10.11. It is possible to overcome this problem by either increasing the clock frequency or increasing the step size.

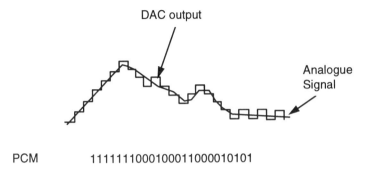

PCM   1111111000100011000010101

Figure 10.9   Delta modulator signal

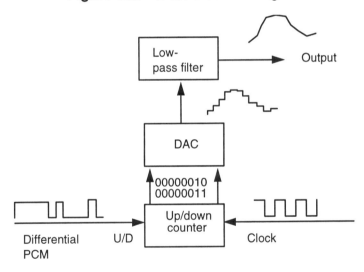

Figure 10.10   Delta modulator receiver

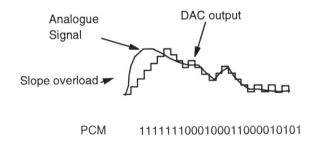

PCM   1111111000100011000010101

Figure 10.11   Slope overload

Granular noise occurs when the signal changes slowly in amplitude, as illustrated in Figure 10.12. The reconstructed signal contains a noise which is not present at the input. Granular noise is equivalent to quantization noise in a PCM system. It can be reduced by decreasing the step size, though there is compromise between this and slope overload.

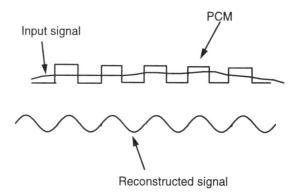

Figure 10.12   Granular noise

10.4.1 Adaptive delta modulation PCM

One method of reducing granular noise and slope overload is to use AΔPCM. With this method the step size is varied by the slope of the input signal. The larger the slope the larger the step size, see Figure 10.13. Algorithms used usually depend on the system and the characteristics of the signal. A typical algorithm is to start with a small step and increase it by a multiple until the required level is reached.

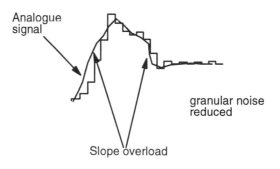

Figure 10.13   Variation of step size

10.5 DIFFERENTIAL PCM (DPCM)

Speech signals tend not to change much between two samples. Thus similar codes are sent, which leads to a degree of redundancy. For example, for a certain sample it is likely that the signal will only change within a range of voltages, as illustrated in Figure 10.14

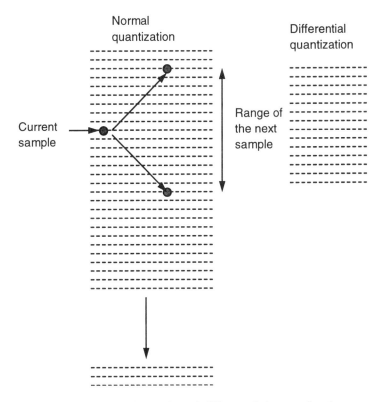

Figure 10.14 Normal and differential quantization

DPCM reduces the redundancy by transmitting the difference in the amplitude of two consecutive samples. Since the range of sample differences is typically less than the range of individual samples, fewer bits are required for DPCM than for conventional PCM.

Figure 10.15 shows a simplified transmitter and receiver. The input signal is filtered to half the sampling rate. This filter signal is then compared with the previous DPCM signal. The difference between them is then coded with the ADC.

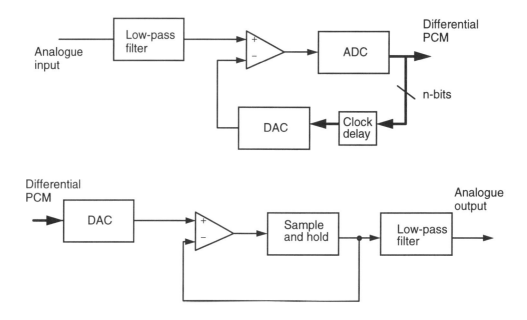

Figure 10.15   DPCM transmitter/receiver

10.5.1 Adaptive differential PCM (ADPCM)

ADPCM allows speech to be transmitted at 32 kbps with little noticeable loss of quality. As with differential PCM the quantizer operates on the difference between the current and previous sample. The adaptive quantizer uses a uniform quantization step, M, but when the signal moves towards the limits of the quantization range the step size, M, is increased. If it is around the centre of the range the step size is decreased. Within any other regions the step size hardly changes. Figure 10.16 illustrates this operation with a signal quanitized to 16-level. This results in 4-bit code.

The change of the quantization step is done by multiplying the quanitization level, M, by a number slightly greater, or less, than 1 depending on the previously quantized level.

10.6 PCM SYSTEMS

PCM is used in ISDN applications and this is discussed in the next chapter. It is also used to transmit multiple speech channels over a single line. This technique uses time-division multiplexing (TDM). In the UK a 30-

channel PCM system is used, whereas the USA uses 24.

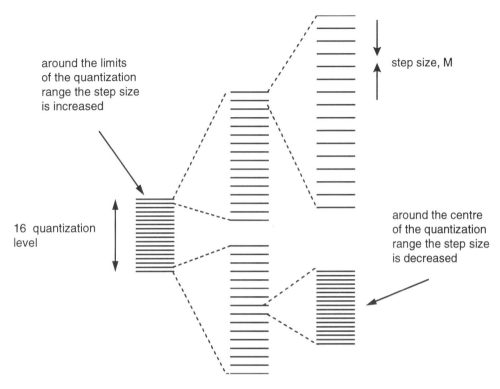

around the limits
of the quantization
range the step size
is increased

step size, M

16 quantization
level

around the centre
of the quantization
range the step size
is decreased

Figure 10.16 ADPCM quantization

With a PCM-TDM system, several voice band channels are sampled, converted to PCM codes, these are then time-division multiplexed onto a single transmission media.

Each channel sampled is given a time slot and these time slots for each of the channels are built up into a frame. The complete frame usually has extra data added to it such as synchronization data, and so on. Speech channels have a maximum frequency content of 4 kHz and are sampled at 8 kHz. This gives a sample time of 125 μs. In the UK a frame is built up with 32 time slots from TS0 to TS31. TS0 and TS16 provide extra frame and synchronization data. Each of the time slots has 8 bits, therefore the overall bit rate is:

Bits per time slot   = 8 bits
Number of time slots = 32
Time for frame   = 125 μs

$$\text{Bit rate} = \frac{\text{No of bits}}{\text{Time}} = \frac{32 \times 8}{125 \times 10^{-6}} = 2048 \text{ kbits / sec}$$

In the USA this bit-rate is 1.544 Mbps. These bit-rates are known as the primary rate multipliers. Further interleaving of several primary rate multiplexes increases the rate to 6.312, 44.736 and 139.264 Mbps (for the USA) and 8.448, 34.368 and 139.264 Mbps (for the UK).

The UK multiframe format is given in Figure 10.17.

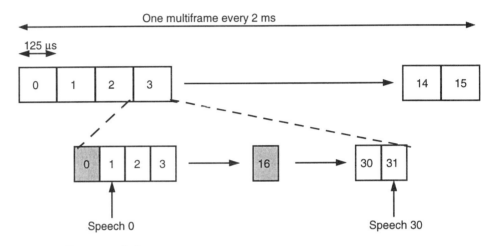

Time slot 0 - Frame word alignment
Time slot 16 - Signalling information

Figure 10.17   30 speech channels PCM-TDM multiframe format

In the UK format the multiframe has 16 frames. Each frame time slot 0 is used for synchronization and time slot 16 for signalling information. This information is sub-multiplexed over the 16 frames. During frame 0 a multiframe-alignment signal is transmitted in TS16 to identify the start of the multiframe structure. In the following frames, the eight binary digits available are shared by channels 1-15 and 16-30 for signalling purposes. TS16 is used as follows:

Frame 0   0000XXXX
Frame 1-15   1234 5678

where 1234 are the four signalling bits for channels 1,2, 3, ..., 15 in consecutive frames, and 5678 are the four signalling bits for channels 16,17,

18 .., 31 in consecutive frames.

Thus in the first frame the 0000XXXX code word is sent, in the next frame the first and the 16th channel appear in TS16, the next will contain the second and the 17th, and on. Typical 4-bit signal information is:

1111 - circuit idle/busy
1101 - disconnection

TS0 contains a frame-alignment signal which enables the receiver to synchronize with the transmitter. The frame-alignment signal (X0011011) is transmitted in alternative frames. In the intermediate frames a signal known as a not-word is transmitted (X10XXXXX). The second binary digit is the complement of the corresponding binary digit in the frame-alignment signal. This reduces the possibility of the demulti- plexed misalignment to imitative frame-alignment signals.

Alternative frames:

TS0:X0011011
TS0:X10XXXXX

where X stands for don't care conditions.

10.7 ALTERNATIVE MARK INVERSION (AMI) LINE CODE

PCM-TDM and ISDN systems use AMI line codes. AMI uses three volt- age levels, where 0 V represents a '0', and the voltage amplitude for each '1' is the inverse of the previous, as shown in Figure 10.18. Note that ISDN uses an inverted AMI code, that is, a '1' is coded as 0 V and a '0' has an invert voltage. A typical pulse amplitude for ISDN is 0.75 V.

The main reason for using the AMI line code is that timing information is inherent in the transmitted signal. This information can be extracted either using a phase-lock loop (PLL) or by averaging the transitions over a relatively long time.

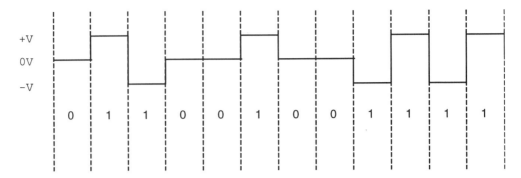

Figure 10.18   AMI coding

Another advantage of AMI line coding is that it reduces the DC power in the signal and its DC voltage content. Electrical power is often sent along a transmission line to provide power for remote circuits where no power supply is available (such as in underwater cables, underground cables or in many telephones). A low-pass filter extracts the DC component from the signal and a high-pass filter recovers the data signal. The DC component is then used to supply power. AMI line coding is an excellent method of coding the bits on the line as it automatically balances the signal voltage and the average voltage will be approximately zero, even, when there are long runs of 0's or 1's.

10.7.1 High density bipolar3 (HDB3) coding

A disadvantage of AMI line coding is that a long run of 0's does not contain timing information, as shown in Figure 10.19.

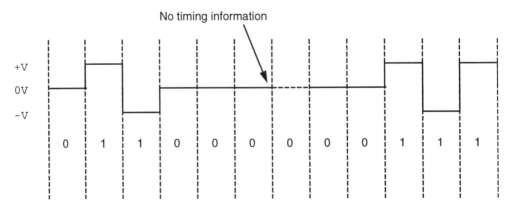

Figure 10.19   Long runs of 0's cause a lack of timing information

This is overcome by using HDB3 coding (in Europe) and B8ZS (in the USA). In AMI if two 1's of the same polarity were transmitted then this would 'violate' the rule of coding. HDB3 uses this to replace any group of four 0's by a sequence of three 0's followed by a violation of the coding rules (i.e. a symbol of the same polarity as the previous). Thus

Bit pattern   `11000001`
Coding   `+-000-0+` (or the inverse)

This coding may cause problems with long runs of 0's as there will be a DC and low frequency component in the data signal, for example:

Bit pattern   `1100000000000001`
Coding   `+-000-000-000-0+` (or the inverse)

will, in this case, contain more negative voltages than positive voltages. To overcome this, a group of consecutive group eight consecutive 0's are coded as `B00VB00V`, where `B` is not a violation and `V` is a volition.

Bit pattern   `100001000011000000001`
Bit pattern   `011110111100111111110`
Coding   `+000+-000-0+-+00+-00-+` (or the inverse)

B8ZS coding is similar but that eight consecutive 1's is coded as `00B0VB0V`.

10.8 TUTORIAL

10.1 Determine the Nyquist sample rate for hi-fi quality music.

10.2 Determine the dynamic range for a 10-bit signed-magnitude PCM code.

10.3 Determine the minimum number of bits required in a PCM code for a dynamic range of 80 dB.

10.4 Determine the time period for a single bit transmission for the UK

PCM-TDM system.

10.5   The USA PCM-TDM system uses 24-channel system. In every sixth frame the least significant bit is 'stolen' for signalling information relating to a particular channel.

(i)   Determine the basic bit-rate.

(ii)   An extra bit is added to the beginning of the 24-frame transmission for synchronization. Determine the bit-rate with the extra synchronization bit.

10.6   Determine the time period for a single bit transmission in the USA PCM-TDM system.

10.7   Determine the signal-to-noise ratio of a 2 V rms signal and a rms quantization noise of 0.2 V.

11

Integrated Services Digital Network (ISDN)

11.1 INTRODUCTION

Chapter 10 shows that to encode speech digitally requires a bit-rate of 64 kbps. Most public telephone networks have now changed from analogue transmission to digital exchanges using PCM switching and transmission. The interface to many telephones is still analogue, as shown in Figure 11.1(a). Over the coming years this will change so that the speech is converted to digital data within the consumer's equipment, as illustrated in Figure 11.1(b). The consumer's equipment will also provide signalling information.

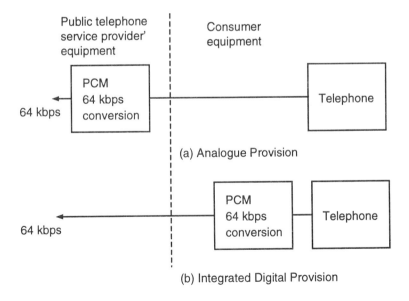

Figure 11.1   Analogue and Digital provision of speech

The great advantage of an ISDN connection is that the type of data transmitted is irrelevent to the transmission and switching circuitry. Thus it can carry other types of digital data, such as facsimile, teletex, videotex and computer data. This reduces the need for modems, which convert digital data into an analogue form, only for the public telephone network to convert the analogue signal back into a digital form for transmission over a digital link.

Figure 11.2 shows the different types of equipment connecting to the ISDN, through network termination equipment (NTE). It is also possible to multiplex the basic rate of 64 kbps to give even higher data rates. This multiplexing is known as N × 64 kbps or Broadband ISDN (B-ISDN).

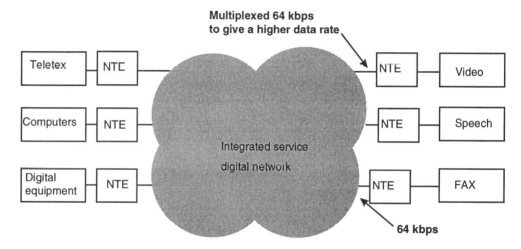

Figure 11.2   ISDN connections

Table 11.1 gives a summary of the CCITT I. standards. These allow universal access to ISDN equipment.

Table 11.1   CCITT standards on ISDN

CCITT standard number	*Description*
I.1*XX*	ISDN terms and technology
I.2*XX*	ISDN services
I.3*XX*	ISDN addressing
I.430 and I.431	ISDN physical layer interface
I.440 and I.441	ISDN data layer interface
I.450 and I.451	ISDN network layer interface
I.5*XX*	ISDN internetworking
I.6*XX*	ISDN maintenance

11.2 ISDN CHANNELS

ISDN uses channels to identify the data rate, each based on the 64 kbps provision. Typical channels are B, D, H0, H11 and H12. The B channel has a data rate of 64 kbps and provides a circuit switching connection between endpoints. The D channel operates at 16 kbps and it controls the data transfers over the B channels. The other channels provide B-ISDN for much higher data rates. Table 11.2 outlines the basic data rates for these channels.

Table 11.2 ISDN channels

Channel	Description
B	64 kbps
D	16 kbps signalling for channel B (ISDN)
	64 kbps signalling for channel B (B-ISDN)
H0	384 kbps (6×64 kbps) for B-ISDN
H11	1.536 Mbps (24×64 kbps) for B-ISDN
H12	1.920 Mbps (30×64 kbps) for B-ISDN

The two main types of interfaces are the basic rate access and the primary rate access. These are both based around groupings of B- and D-channels. The basic rate access allows two B and one 16 kbps D-channel.

Primary rate provides B-ISDN, such as H12 which gives $30 \times B$ channels and a 64 kbps D channel. For basic and primary rates, all channels multiplex onto a single line by combining channels into frames and adding extra synchronization bits. Figure 11.3 gives examples of the basic rate and primary rate.

The basic rate ISDN gives two B-channels at 64 kbps and a signalling channel at 16 kbps. These multiplex into a frame and, after adding extra framing bits, the total output data rate is 192 kbps. The total data rate for the basic rate service is thus 128 kbps. One or many devices may multiplex their data, such as two devices transmitting at 64 kbps, a single device multiplexing its 128 kbps data over two channels (giving 128 kbps), or by several devices transmitting sub-64 kbps data rate over the two channels. For example, four 32 kbps devices could simultaneously transmit their data, or even eight 16 kbps devices, and so on.

For H12, 30×64 kbps channels multiplex with a 64 kbps signalling channel, and with extra framing bits, the resulting data rate is 2.048 Mbps (compatible with European PCM-TDM systems). The actual data rate is 1.920 Mbps. As with the basic service this could contain a number of devices with a data rate of less than or greater than a multiple of 64 kbps.

For H11, 24 × 64 kbps channels multiplex with a 64 kbps signalling channel, and with extra framing bits, results in a data rate of 1.544 Mbps (compatible with USA PCM-TDM systems). The actual data rate is 1.536 Mbps.

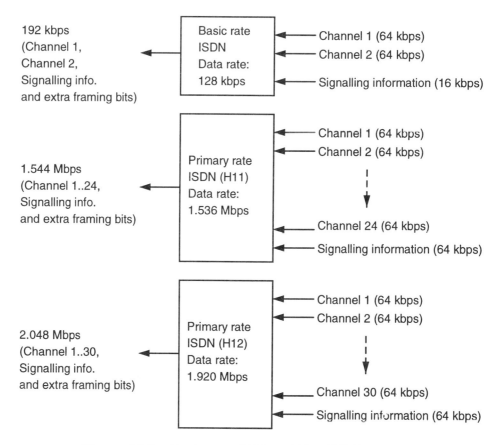

Figure 11.3   Basic rate, H11 and H12 ISDN services

11.3 ISDN PHYSICAL LAYER INTERFACING

The physical layer corresponds to layer 1 of the OSI 7-layered model and is defined in CCITT specifications I.430 and I.431. Pulses on the line are not coded as pure binary, they use a technique called Alternate Mark Inversion (AMI).

11.3.1 Alternative Mark Inversion (AMI) line code

The AMI line code uses three voltage levels. In pure AMI, 0 V represents a '0', and the voltage amplitude for each '1' is the inverse of the previous. ISDN uses the inverse of this, that is, 0 V for a '1' and an inverse in voltage for a '0', as shown in Figure 11.4. Normally the pulse amplitude is 0.75 V.

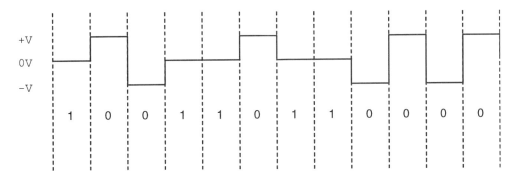

Figure 11.4   AMI used in ISDN

The inversion the AMI signal (that is, inverting a '0' rather than the '1') allows for timing information to be recovered when there are long runs of zeros, which is typical in the idle state. AMI line also automatically balances the signal voltage and the average voltage will be approximately zero, even, when there are long runs of 0's.

11.3.2 System connections

In basic rate connections up to eight devices, or termination equipment (TE), can connect to the network termination (NT), as shown in Figure 11.5. These connect over a common four wire bus using two sets of twisted-pair cables. The transmit output (T_x) on each TE connects to the transmit output on the other TEs, and the receive input (R_x) on each TE connects to all other TEs. On the NT the receive input connects to the transmit of the TEs and the transmit output of the NT connects to the receive of the TEs. A contention protocol allows only one TE to communicate at a time.

An on-board DC power supply normally provides power for the TEs. If this were to fail then, in emergencies, the TE can use power supplied from the NT. Figure 11.6 shows how the NT provides power either through the primary supply, using the T_x/R_x lines or through two other dedicated lines. A centre-tapped transformer provides the primary supply of power and gives a nominal voltage of 40 V at 1 W.

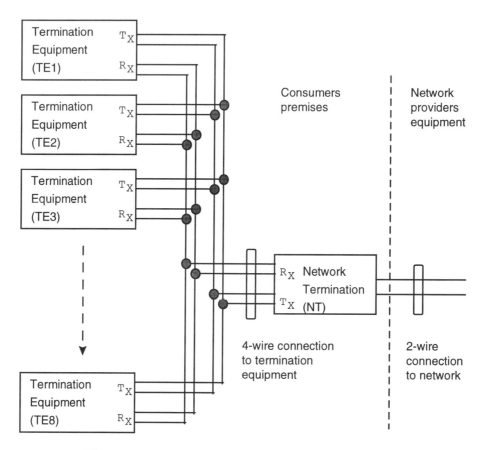

Figure 11.5   Connection of termination equipment

An 8-pin ISO 8877 connector connects a TE to the NT, this is similar to the RJ-45 connector but has two extra pin connections. Pins 3 and 6 carry the T_x signal from the TE, pins 4 and 5 provide the R_x to the TEs. Pins 7 and 8 are the secondary power supply from the NT and pins 1 and 2 the power supply from the TE (if used). The T_x/R_x lines connect via transformers thus only the AC part of the bit stream transfers into the PCM circuitry of the TE and the NT. Thus need for a balanced DC line code such as AMI, as the DC component in the bit stream will not pass through the transformers.

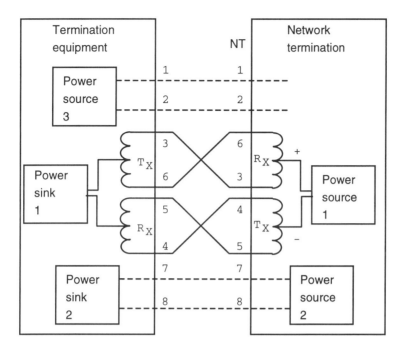

Figure 11.6 Power supplies between NT and TE

11.3.3 Frame format

Figure 11.8 and 11.9 show the ISDN frame formats. Each frame is 250 μs long and contains 48 bits, this give a total bit rate of 192 kbps ($48/250 \times 10^{-6}$) made up of two 64 kbps B channels, one 16 kbps D-channel and extra framing, DC balancing and synchronization bits.

The frame format for TE to NT transmission differs from an NT to a TE transmission. Figure 11.8 shows the format for TE to NT and Figure 11.9 shows the NT to TE format. The F/L pair of bits identify the start of each transmitted frame.

When transmitting from a TE to an NT there is a 10-bit offset in the return of the frame back to the TE, as shown in Figure 11.7. The E bits echo the D-channel bits back to the TE.

When transmitting from the NT to the TE, the bits after the F/L bits, in the B-channel, has a volition in the first 0. If any of these bits is a 0 then a volition will occur, but if they are 1's then no volition can occur. To overcome this the F_A bit forces a volition. Since it is followed by 0 (the N bit) it will not be confused with the F/L pair. The start of the frame can thus be traced backwards to find the F/L pair.

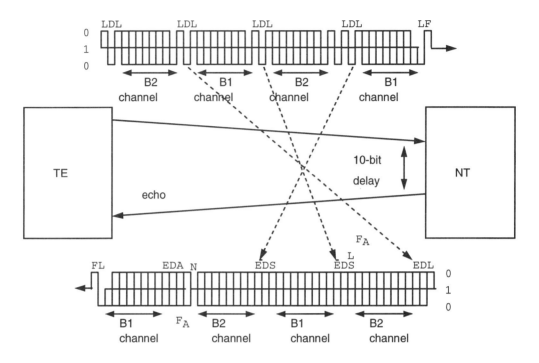

Figure 11.7   Frame transmission and echo when transmitting from TE to NT

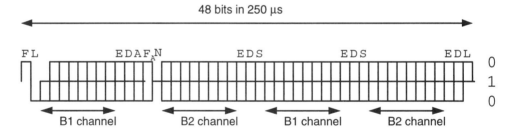

Figure 11.8   ISDN frame format for NT to TE

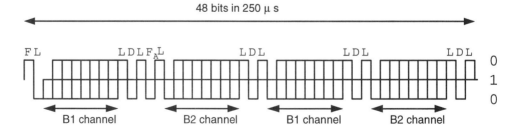

Figure 11.9   ISDN frame format for TE to NT

where F – framing bit   N   – set to a 1
      L – DC balancing bit   D   – D-channel bit
      E – D-echo channel bit F_A – auxiliary framing bit (=0)
      S – reserved for future use A   – activation bit
      M – multiframing bit   B1 – bits for channel 1
     B1 – bits for channel 2

There are 16 bits for each B-channel giving a basic data rate of 64 kbps $(16/250 \times 10^{-6})$ and there are 4 bits in the frame for the D-channel, giving a bit rate of 16 kbps $(4/250 \times 10^{-6})$.

The L bit balances the DC level on the line. If the number of 0's following the last balancing bit is odd then the balancing bit is a 0, else it is a 1.

When synchronized the NT informs the TEs by setting the A bit.

11.3.4 D-channel contention

The D-channel contention protocol ensures that only one terminal can transmit its data at a time. This happens because the start and the end of the D-channel bits has the bit stream 01111110, as shown next:

11111**01111110**XXXXXXXXX......XXXXXXXX**01111110**1111

When idle, each TE floats to a high impedance state, which is taken as a binary 1. To transmit a TE counts the number of 1's in the D-channel. A 0 resets this count. After a predetermined number, greater than a predetermined number of consecutive 1's then the TE transmits its data and monitors the return from the NT. If it does not receive the correct D-channel bit stream returned through the E bits then a collision has occurred. When a TE detects a collision it immediately stops transmitting and monitors the line.

When a TE has finished transmitting data it increases its count value for the number of consecutive 1's by 1. This gives other TEs an opportunity to transmit their data.

11.4 ISDN DATA LINK LAYER

The data link layer uses the HDLC (High level data link control) format, commonly known as the Link Access Procedure for the D-channel (LAPD).

Figure 11.10 shows the frame format. The unique bit sequence 01111110 identifies the start and end of the frame. This bit pattern cannot occur in the rest of the frame due to zero bit-stuffing. With zero bit-stuffing the transmitted inserts a zero into the bit stream when transmitting five consecutive 1's. When the receiver receives five consecutive 1's it deletes the next bit if it is a zero. This stops the unique 01111110 sequence occurring within the frame (as discussed in section 6.3).

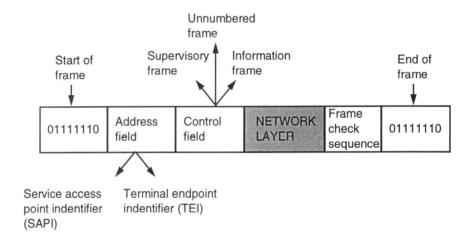

Figure 11.10   D-channel frame structure

The address field contains information on the type of data contained in the frame (the service access point identifier) and the physical address of the ISDN device (the terminal endpoint identifier). The control field contains either a supervisory, unnumbered or an information frame. The frame check sequence provides error detection information.

11.4.1 Address field

The data link address only contains addressing information to connect the TE to the NT and does not have network addresses. Figure 11.11 shows the address field format. The SAPI identifies the type of ISDN services.

For example a frame from a telephone would be identified as such, and only telephones would read the frame.

All TEs connect to a single multiplexed bus, thus each has a unique data link address, known as a Terminal Equipment Identifier (TEI). The user or the network set this, the ranges of available addresses are:

0-63	non-automatic assignment TEIs
64-126	automatic assignment TEIs
127	global TEI

The non-automatic assignment involves the user setting the address of each of the devices connected to the network. When a device transmits data it inserts its own TEI address and only receives data which has its TEI address. In most cases devices should not have the same TEI address as this would cause all devices with the same TEI address, and the SAPI, to receive the same data (although, in some cases, this may be a requirement).

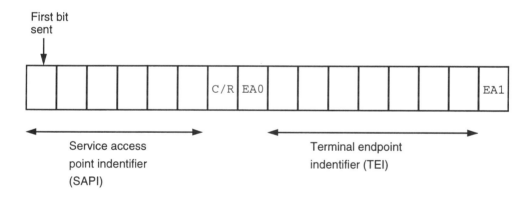

Figure 11.11   Data link address field

The network allocates address to devices requiring automatic assignment before they can communicate with any other devices. The global TEI address is used to broadcast messages to all connected devices. A typical example is when a telephone call is incoming to a group on a shared line where all the telephones would ring until one was answered.

The C/R bit is the command/ response bit and EA0/EA1 are extended address field bits.

11.4.2 Control field

ISDN uses the asynchronous balanced mode extended (ABME). This mode uses a 16-bit control field for information and supervisory frames and an 8-bit field for unnumbered frames. Figure 11.12 shows the extended control field. Chapter 6 contains a fuller description of this field.

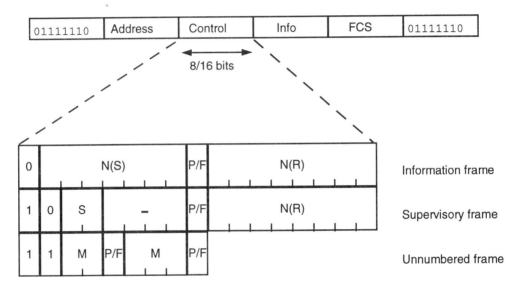

Figure 11.12 ISDN control field

Information frames contain sequenced data. The format is 0SSSSSSSXRRRRRRR, where SSSSSSS is the send sequence number and RRRRRRR is the frame sequence number that the send expects to receive next (X is the poll/final bit). Since the extended mode uses a 7-bit sequence field then information frames are numbered from 0 to 127.

Supervisory frames contain flow control data. Table 11.3 lists the supervisory frame types and the control field bit settings. The RRRRRRR value represent the 7-bit receive sequence number.

Table 11.3 Supervisory frame types and control field settings

Type	Control field setting
Receiver Ready (RR)	10000000PRRRRRRR
Receiver Not Ready (RNR)	10100000PRRRRRRR
Reject (REJ)	10010000PRRRRRRR

Unnumbered frames set-up and clear connections between a node and the network.. Table 11.4 lists the unnumbered frame commands and Table 11.5 lists the unnumbered frame responses.

Table 11.4   Unnumbered frame command and control field settings

Type	*Control field setting*
Set asynchronous balance mode extended (SABME)	1111P110
Unnumbered Information (UI)	1100F000
Disconnect mode (DISC)	1100P010

Table 11.5   Unnumbered frame responses and control field settings

Type	*Control field setting*
Disconnect mode (DM)	1111P110
Unnumbered acknowledgement (UA)	1100F000
Frame reject (FRMR)	1110P001

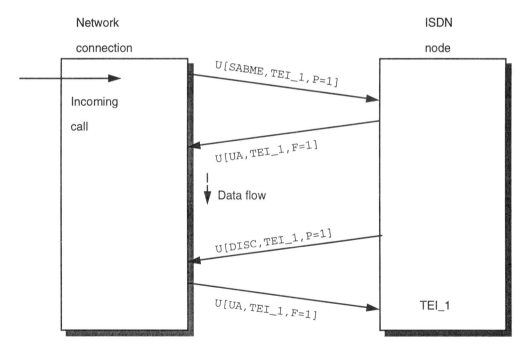

Figure 11.13   Connection between a primary/secondary in SABME

As ISDN uses set asynchronous balanced mode extended (SABME)

then all connected nodes and the network connection can send commands and receive responses but the network connection always acts as the primary. Figure 11.13 shows a sample connection of an incoming call to an ISDN node (address `TEI_1`). The SABME mode is set-up initially using the SABME command (`U[SABME,TEI_1,P=1]`). followed by an acknowledgement from the ISDN node (`U[UA,TEI_1,F=1]`). At any time, either the network or the node can disconnect the connection. In this case the ISDN node disconnects the connection with the command `U[DISC,TEI_1,P=1]`. The network connection acknowledges this with an unnumbered acknowledgement (`U[UA,TEI_1,F=1]`).

In the example in Figure 11.13 there is an incoming call. If the call was outgoing then the ISDN node would initiate the SABME command and the acknowledgement would come from the network connection.

Once the connection has been made then information frames can flow between the nodes. The flow is controlled by the supervisory frames. The number of frames sent before an acknowledgement is required is set by the window size. This can vary from 1 to 127 and normally depends on the type of data transmitted. With telephone-type data the window size is 1. This means that an acknowledgement must be sent for every frame received and the sender of the data will not send any more data until it has received the acknowledgement. Refer to section 6.4.2 for examples of data flow. Note that that frames will have the ISDN data link address as there can be many secondary nodes (ISDN nodes) but only one primary node (the network connection).

11.4.3 Frame check sequence

The frame check sequence (FCS) field contains an error detection code based on cyclic redundancy check (CRC) polynomials. It uses the CCITT V.41 polynomial, which is $G(x) = x^{16} + x^{12} + x^5 + x^1$.

11.5 ISDN NETWORK LAYER

The D-channel carriers network layer information within the LAPD frame. This information establishes and controls a connection. The LAPD frames contain no true data as this is carried in the B-channel. Its function is to sets up and manage calls and to provides flow control between connections over the network. In addition, it is possible to use the D-channel as a carrier for a conventional packet-switched network based around the X.25 protocol. A schematic of how the layers interact is shown in Figure 11.14. The description is this section relates mainly to the control of circuit-switched calls.

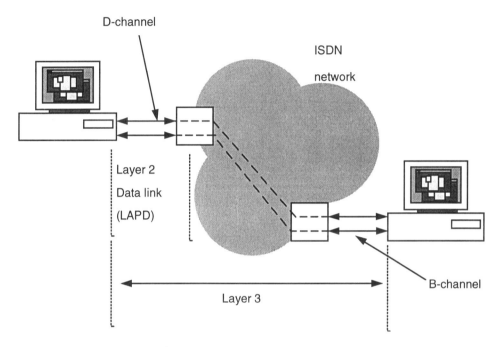

Figure 11.14   Connection of two nodes to an ISDN network

Figure 11.15 shows the format of the layer 3 signalling message frame. The first byte is the protocol discriminator. In the future this byte will define different communications protocols. At present it is normally set to `0001000`. After the second byte the call reference value is defined. This is used to identify particular calls with a reference number. The length of the call reference value is defined within the second byte. As it contains a 4-bit value, up to 16 bytes can be contained in the call reference value field. The next byte gives the message type and this types define the information contained in the proceeding field.

There are four main types of messages, these are call establish, call information, call clearing and miscellaneous messages. The message types for the call establish messages are:

ALERTING	00000001
CALL PROCEEDING	00000010
CONNECT	00000111
CONNECT ACKNOWLEDGE	00001111
PROGRESS	00000011
SETUP	00000101
SETUP ACKNOWLEDGE	00001101

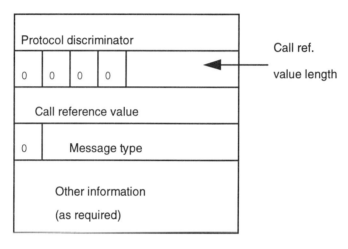

Figure 11.15   Signalling message structure

The message types for the call information phase message are:

RESUME	00100110
RESUME ACKNOWLEDGE	00101110
RESUME REJECT	00100010
SUSPEND	00100101
SUSPEND ACKNOWLEDGE	00101101
SUSPEND REJECT	00100001
USER INFORMATION	00100000

The message types for the call clearing messages are:

DISCONNECT	01000101
RELEASE	01001101
RELEASE COMPLETE	01011010
RESTART	01000110

RESTART ACKNOWLEDGE   01001110

Other miscellaneous message types are:

SEGMENT	01100000
CONGESTION CONTROL	01111001
INFORMATION	01111011
FACILITY	01100010
NOTIFY	01101110
STATUS	01111101
STATUS ENQUIRY	01110101

Figure 11.16 shows an example connection procedure. The initial message sent is SETUP. This may contain some of the following:

- channel identification - identify a channel with an ISDN interface;
- calling party number;
- calling party sub-address;
- called party number;
- called party sub-number;
- extra data (2-131 bytes).

After the calling TE has sent the SETUP message. The network then returns the SETUP ACK message. If there is insufficient information in the SETUP message then other information needs to flow between the called TE and the network. After this the network sends back a CALL PROCEEDING message and it also sends a SETUP message to the called TE. When the called TE detects its TEI address and SAPI, it and sends back an ALERTING message. This informs the network that the node is alerting the user to answer the call. When it is answered then the called TE sends a CONNECT to the network. The network then acknowledges this with a CONNECT ACK message, at the same time it sends a CONNECT message to the calling TE. The calling TE then acknowledges this with a CONNECT ACK. The connection is then established between the two nodes and data can be transferred.

To disconnect the connection the DISCONNECT, RELEASE and RELEASE COMPLETE messages are used.

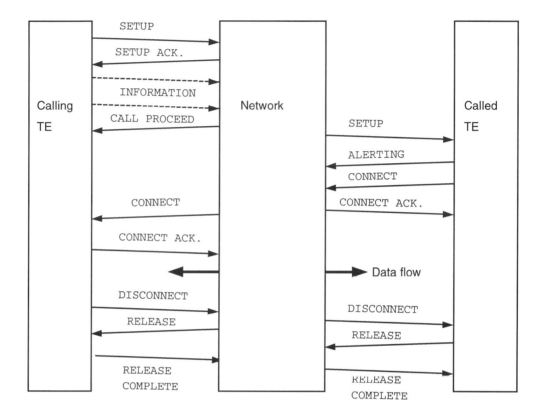

Figure 11.16   Call establishment and clearing

11.6 LOCAL AREA NETWORKS CONNECTED TO ISDN NETWORKS

ISDN proves a good reliable, well proven, method of connecting LANs over a relatively large distance. Figure 11.17 shows some ways which nodes connect over an ISDN network.

An ISDN bridge/router routes frames from a LAN into two B-channels and a single D-channel, for transmission over the ISDN. The router is typically either a network server node with an ISDN plug-in card inserted into it, or is a dedicated router. It normally also contains hardware to compress video, speech, audio and computer data to reduce the amount of data transmitted. A typical ISDN LAN-LAN bridge/router can transmit WAN data, over a leased line using B-ISDN, at, up to, 2 Mbps and supports TCP/IP and SPX/IPX protocols.

ISDN is also useful for connecting remote computers with ISDN interfaces, as shown in Figure 11.17. This is typically used in remote office locations or home working. The connection to the ISDN network is either with an RJ-45 connector or an RS-232 25-pin connector.

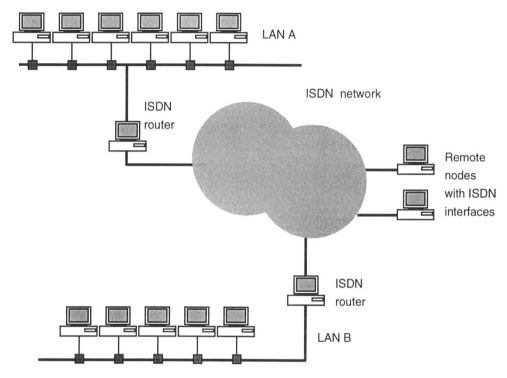

Figure 11.17 ISDN connections between LANs

11.7 TUTORIAL

11.1 Show why speech requires to be transmitted at 64 kbps.

11.2 If the bandwidth of Hi-Fi audio is 20 kHz and 16 bits are used to code each sample, determine the required bit rate for single channel transmission.

11.3 Explain the format of the ISDN frame.

11.4 If an ISDN frame has 48 bits and takes 250 μs to transmit. Show that the bit rate on each D-channel is 16 kbps and that the bit rate of the B-channel is 64 kbps.

11.5   Explain the different types of frames and show how a connection is made between to ISDN nodes.

11.6   Show how the supervisory frames are used to control the flow of data.

11.7   Discuss the format of the ISDN network layer packet.

11.8   Discuss how an ISDN node sets up and disconnects a network connection.

12

Asynchronous Transfer Mode (ATM)

12.1 INTRODUCTION

Asynchronous transfer mode (ATM) has been developed mainly by the telecommunications companies and is proposed as a standard for Broadband ISDN. It was independently developed by Bellcore, the research arm of AT&T in the USA, and several giant telecommunications companies in Europe. This has led to two possible future standards.

In the USA the ANSI T1S1 subcommittee have supported and investigated ATM and in Europe it has been investigated by ETSI. There are small differences between the two proposed standards, but these may converge into one common standard. The CCITT has also dedicated a study group XVIII to Broadband ISDN with the objective of merging differences and creating a single global world wide standard for broadband networks.

A major objective of telecommunications companies is to integrate real-time data (such as voice and video information) and non real-time data (such as computer data and file transfer). Computer data can, typically, be transferred in non real-time but it is important that the connection is free of errors. A single bit error on computer data can cause serious damage. Voice and video data, on the other hand, require a constant sampling rate and low propagation delays, but they are more tolerant to errors and losses of small parts of the data. A mechanism for providing for real-time and non real-time data helps integrate data communications services.

The main problem with carrying these different types of traffic is that real-time data requires different sampling, or scanning, times, depending on the application. For example a high-resolution video image may need to be sent as several megabytes of data in short time burst, but then nothing for a few seconds.

For voice data the signal must be sampled 8 000 times per second. This is the reason that ISDN and PCM use this rate as a base transmission. Unfortuately few other applications require either constant sampling or this sample rate.

Computer data will, typically, be sent in bursts. Sometimes a high transfer rate is required (for example when running a computer package remotely over a network) or a relatively slow transfer (such as when reading text information). ISDN and PCM-TDM are thus wasteful because they either allocate a switched circuit (for ISDN) or reserve fixed time slot (in PCM-TDM), no matter if there is data being transmitted at that time. It may not be possible to service high burst rate by allocating either time slots or switched circuits because all of the other time slots are full, or switched circuits are being used.

ISDN and PCM-TDM use a synchronous transfer mode (STM) technique where a connection is made between two devices by circuit switching. The transmitting device is assigned a given time slot to transmit their data. This timeslot is fixed for the period of the transmission. The main problems with this type of transmission are:

- that not all the time slots are filled by data when there is light data traffic. This is wasteful in data transfer;
- when a connection is made between two end points a fixed time slot is assigned and data from that connection is always carried in that time slot. This is also wasteful because there may be no data being transmitted in certain time periods.

ATM overcomes these problems by splitting the data up into small fixed-length packets, known as cells. Each data cell is sent with its connection address and follow a fixed route through the network. The packets are small enough that, if they are lost, due possibly to congestion, they can either be requested (for high reliability) or cause little degradation of the signal (typically in voice and video traffic).

This is very similar to packet switching and uses fast packet switching with short fixed packet lengths. The short packet lengths allow for little degradation in voice transmission when the occasional packet is lost in the network.

The address of devices on an ATM network are identifier by a Virtual Circuit Identifier (VCI), instead of by a time slot as in a STM network. The VCI is carried in the header portion of the fast packet.

The packets are of fixed length and to differentiate then from X.25 variable length packets they are normally referred to as cells.

12.2 STATISTICAL MULTIPLEXING

Fast packet switching attempts to solve the unused the time slot problem of STM. This is achieved by statistically multiplexing several connections on the same link based on their traffic characteristics. Applications, such as voice traffic, which require a constant data transfer are allowed safe routes through the network. Whereas several applications, which have bursts of traffic, may be assigned to the same link in the hope that statistically they will not all generate bursts of data at the same time. Even if some of them were to burst simultaneously, then their data could be buffered and sent at a later time. This technique is called statistical multiplexing and allows the average traffic on the network to be evened-out over a relatively short time period. This is impossible on an STM network.

Considerable amount of research has been done on statistical multiplexing of real-time versus non real-time data, and burst data against continuous data. This is not a major problem under light traffic loads, but under heavy loads an ATM system degenerates into an STM-type system.

12.3 ATM USER NETWORK INTERFACES (UNI)

Users access an ATM network through a user network interface (UNI). This transmits data into the network with a set of agreed specifications. The network must then try and ensure that the connection stays within those requirements and that it satifies the required quality of service for the entire duration of the connection.

It is likely that there will be several different types of ATM service provision. One type will provide an interface to one or more of the LAN standards (such as Ethernet or token ring) or FDDI. The conversion of the LAN frames to ATM cells will be done inside the UNI at the source and destination endpoints respectively. Figure 12.1 shows the gateway connecting the Ethernet, token ring, FDDI, or other LAN/MAN interface to the UNI. Typically it will be used as a bridge for two widely separated LANs. This provides a short term solution to justifying the current investment in LAN technology and allows a gradual transition to complete ISDN/ ATM networks.

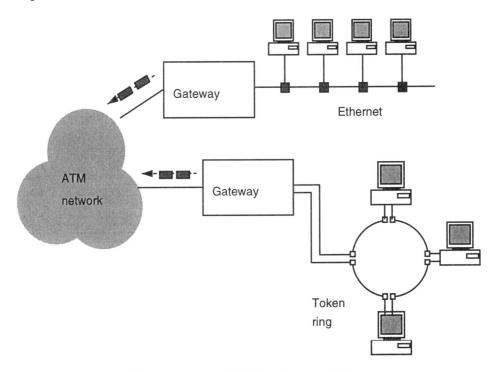

Figure 12.1   ATM interface to LANs

The best long-term solution is to connect data communication equipment directly onto an ATM network. This allows computer equipment, telephones, video, etc. to connect directly to a global network.The output from an ATM multiplexer interfaces the UNI of a larger ATM backbone network. This is illustrated in Figure 12.2.

A third different type of ATM interface connects existing STM networks to ATM networks. This allows a slow migration of existing STM technology to ATM.

The IEEE 802.6 standard for the MAC layer of the Metropolitan Area Network (MAN) DQDB (Distributed Queue Dual Bus) protocol is similar to the ATM cell.

12.4 ATM CELLS

The ATM cell, as specified by ANSI T1S1 sub-committee, has 53 bytes, as shown in Figure 12.3. The first 5 bytes is the header and the remainder can hold 48 bytes of data. Optionally the data can contain a 4 byte ATM adaptation layer and 44 bytes of actual data. A bit in the control field of the header set either 44 or 48 bytes of data. The ATM adaptation layer field allows for fragmentation and reassembly of cells into larger packets at the source and destination respectively. The control field also contains bits which specify whether this is a flow control cell or an ordinary data cell, a bit to indicate if this packet can be deleted in a congested network, and so on.

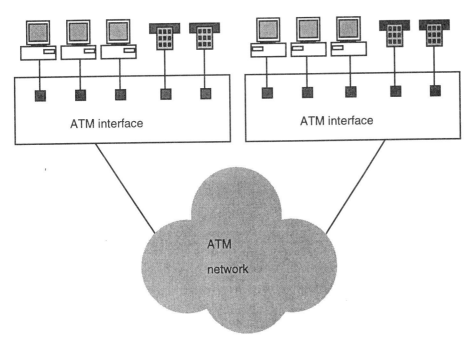

Figure 12.2   Direct connection of data communication equipment to ATM

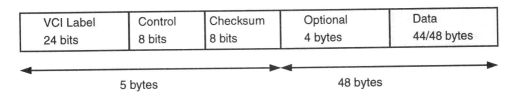

VCI Label 24 bits	Control 8 bits	Checksum 8 bits	Optional 4 bytes	Data 44/48 bytes

5 bytes   48 bytes

Figure 12.3   ATM cell

The ETSI definition of an ATM cell also contains 53 bytes with a 5 byte header and 48 bytes of data. The main differences are the number of bits in the VCI field, the number of bits in the header checksum, and the definitions and position of the control bits.

12.5 CONNECTIONS ON AN ATM NETWORK

In STM networks, data can change its position in each time slot in the interchanges over the global network. This can occur in ATM where the VCI label changes between intermediate nodes in the route.

When a transmitting node wishes to communicate through the network it makes contact with the UNI and negotiates parameters such as destination, traffic type, peak and traffic requirements, delay and cell loss requirement, and so on. The UNI forwards this request to the network. From this data the network computes a route based on the specified parameters and determines which links on each leg of the route can best support the requested quality of service and data traffic. It sends a connection set-up request to all the nodes in the path enroute to the destination node. The routing is similar to the set-up of a virtual circuit in X.25 packet switching. Refer to Figure 7.12 for an example virtual connection and section 7.5 for more information. It differs from X.25 in that no acknowledgements are sent back for data received. ATM also differs from X.25 in that the data cells do not have sequence numbers.

One the connection has been made, data can be transmitted with the required quality of service. When the connection is terminated the VCI labels assigned to the communications are used for other connections.

Certain users, or applications, can be assigned reserved VCI labels for special services that may be provided by the network. However, as the address field only has 24 bits it is unlikely that many of these requests would be granted.

Note that as there is a virtual circuit set-up between the transmitting and receiving node then cells are always delivered in the order they were transmitted. This is because cells cannot take alternative routes to the destination. Even if the cells are buffered at a node then they will still be transmitted in the correct sequence.

It is possible to send cells into the network without the use of virtual circuits by using the adaptation layer. This may cause cells to arrive at the receiver out-of-sequence. The layer thus contains a sequence number to allow for future reassembly at the receiver.

12.6 ATM AND THE OSI MODEL

The basic ATM cell fits roughly into the data link layer of the OSI model, but contains some network functions, such as end-to-end connection, flow control, and routing. It thus fits into layers 2 and 3 of the model, as shown in Figure 12.5. The layer 4 software layer, such as TCP/IP, can communicate directly with ATM.

The ATM network provides a virtual connection between two gateways and the IP protocol fragments IP packets into ATM cells at the transmitting UNI which are then are reassembled into the IP packet at the destination UNI.

With TCP/IP each host is assigned an IP address as is the ATM gateway. Once the connection has been made then the cells are fragmented into the ATM network and follow a predetermined route through the network. At the receiver the cells are reassembled using the ATM adaptation layer. This then reforms the original IP packet which is then passed to the next layer.

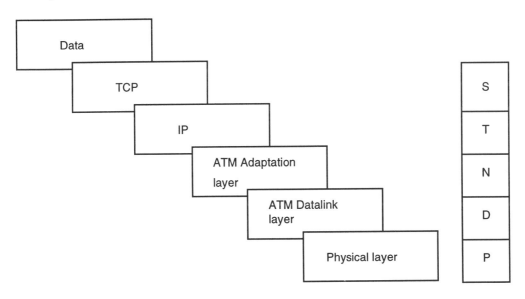

Figure 12.4 ATM and the OSI model

12.7 ATM PHYSICAL LAYER

The physical layer is not an explicitly part of the ATM definition, but is currently being considered by the standards organizations. T1S1 has standardized on SONET (Synchronous Optical NETwork) as the preferred physical layer, with STS-3c at 155.5 Mbps, STS-12 at 622 Mbps and STS-48 at 2.4 Gbps.

The SONET physical layer specification provides a standard world wide digital telecommunications network hierarchy, known internationally as the Synchronous Digital Hierarchy (SDH). The base transmission rate, STS-1, is 51.84 Mbps. This is then multiplexed to make up higher bit rate streams, such as STS-3 which is 3 times STS-1, STS-12 which is 12 times STS-1, and so on. The 155 Mbps stream is the lowest bit rate for ATM traffic and is also referred to as STM-1 (Synchronous Transport Module - Level 1).

The SDH specifies a standard method on how data is framed and transported synchronously across fibre optic transmission links without requiring that all links and nodes have the same synchronized clock for data transmission and recovery.

12.8 ATM FLOW CONTROL

ATM cannot provide for a reactive end-to-end flow control because by the time a message is returned from the destination to the source, large amounts of data could have been sent along the ATM pipe, possibly making the congestion worse. The opposite can occur when the congestion on the network has cleared by the time the flow control message reaches the transmitter. The transmitter would thus reduce the data flow when there is little need. ATM tries to react to network congestion quickly, and slowly reduce the input data flow to reduce congestion.

This rate-based scheme of flow control involves controlling the amount of data to a specified rate that is agreed upon when the connection is made. It then automatically changes the rate based on the past history of the connection as well as the present congestion state of the network.

Data input is thus controlled by detecting traffic congestion early by closely monitoring the internal queues inside the ATM switches, as shown in Figure 12.5. The network then reacts gradually as the queues lengthen and reduces the traffic into the network from the transmitting UNI. This is an improvement over imposing a complete restriction on the data input when the route is totally congested. In summary, anticipation is better than desperation.

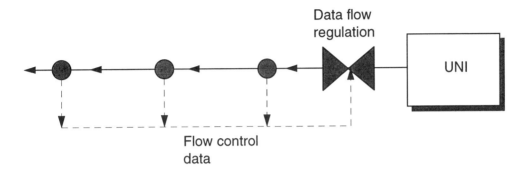

Figure 12.5   Flow control feedback from ATM switches

A major objective of flow control scheme is to try and only affect the streams which are causing the congestion, and not affect well behaved streams.

12.9 RELIABILITY OF AN ATM NETWORK

There is no end-to-end reliable checking service in ATM as it does not request retransmissions and there are no end-to-end acknowledgements for received data. TCP/IP (or any layer above ATM) can provide these services.

12.10 ATM PERFORMANCE

ATM networks rely on user supplied information to profile traffic flows so that a connection has the desired service quality. Some traffic sources are easier to characterize than others. Table 12.1 gives four basic data types.

Table 12.1   Four basic categories of data

Loss sensitive	Delay sensitive	Example data type
yes	yes	Real-time control system
no	yes	Telephone/hi-fi music
yes	no	File transfer, application programs
no	no	Teletex information

It is further complicated by differing data types either sending data in a continually repeating fashion (such as telephone data) or with a variable frequency (such as interactive video).

12.11 PRACTICAL ATM NETWORKS

Many cities in the UK are investing in Metropolitan Area Networks which use ATM and/or FDDI technology. In Scotland, several academic institutions have installed MANs around their local area, these include:

- EaStMAN (Edinburgh/ Stirling);
- ClydeNet (Glasgow area);
- FaTMAN (Fife and Tayside); and,
- AbMAN (Aberdeen area).

In Edinburgh and Stirling the EaStMAN phase I network connects ten sites in Edinburgh and also the University of Stirling, as shown in Figure 12.6. The Edinburgh network connects to the major campuses of the University of Edinburgh (UoE), Heriot-Watt University, Napier University, Edinburgh College of Art, Moray House Institute of Education and Queen Margaret College (QMC). This then links to other academic networks through the high speed SuperJANET (Joint Academic NETwork).

The two rings of FDDI and ATM have been run around the Edinburgh based sites. This also connects to the University of Stirling through a 155 Mbps SDH connection. Two connections on the ring are made to the SuperJANET network. These are at the Heriot-Watt University and the University of Edinburgh.

The 100 Mbps FDDI dual rings links 10 Edinburgh city sites. This ring provides for IP traffic on SuperJANET and also for high speed metropolitan connections. Initially a 155 Mbps SDH/STM-1 ATM network connects five Edinburgh sites and the University of Stirling. This also connects to the SuperJANET ATM pilot network. Figure 12.7 shows the FDDI and ATM connections.

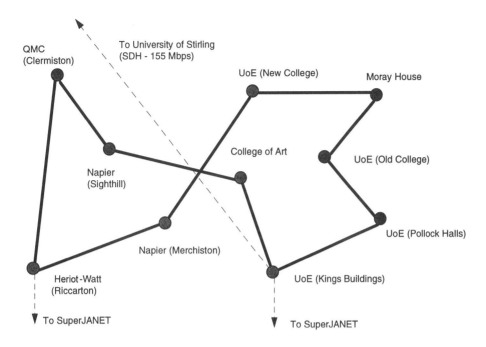

Figure 12.6   EaStMAN phase 1 connections

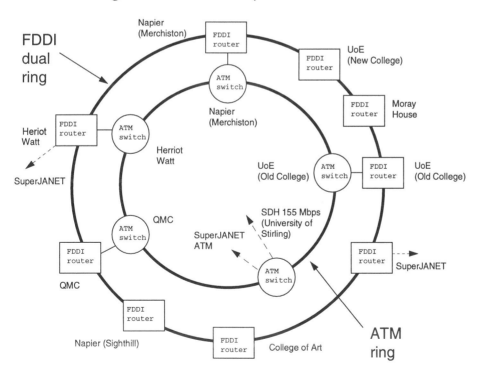

Figure 12.7   EaStMAN ring connections

The two different network technologies allow the universities to operate a two-speed network. For computer-type data the well-established FDDI technology provides good reliable communications and the ATM network allows for future exploitation of mixed voice, data and video transmissions. Future plans are to connect other sites, remote sites and to connect schools, colleges and private and public sector organizations to the rings.

The JANET and SuperJANET network provides a connection to all UK universities. A gateway out of the network to the rest of the world is located at University College London (UCL), as shown in Figure 12.8.

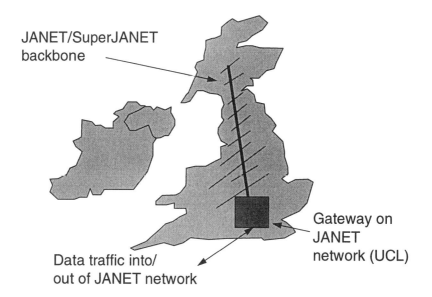

Figure 12.8   JANET/SuperJANET backbone

12.11.1 EaStMAN addressing and routing

Each FDDI interface on the ring has a unique IP address and DNS name, the routers have the following IP addresses and DNS names:

```
Ed Old College      194.81.56.65     oc2.ed.eastman.net.uk
Ed Pollock          194.81.56.66     ph2.ed.eastman.net.uk
Ed New College      194.81.56.67     nc2.ed.eastman.net.uk
Moray House         194.81.56.78     gw1.mhie.eastman.net.uk
College of Art      194.81.56.81     gw1.eca.eastman.net.uk
QMC                 194.81.56.94     gw1.qmced.eastman.net.uk
Napier Merchiston   194.81.56.97     me1.napier.eastman.net.uk
Napier Sighthill    194.81.56.98     si1.napier.eastman.net.uk
```

```
Heriot-Watt      194.81.56.110     gw1.hw.eastman.net.uk
SuperJANET       194.81.56.126     gw1.sj.eastman.net.uk
```

The IP address use a class C network with a netmask of `255.255.255.0`.

12.12 TUTORIAL

12.1   Discuss how an X.25 packet differs from an ATM cell.

12.2   Discuss how an X.25 packet flow differs from an ATM cells flow.

12.3   Discuss how ATM connections are more efficient in their transmission than ISDN and PCM-TDM.

12.4   Investigate a local WAN and determine the network topology and its network technology.

12.5   If there is access to an Internet connection, use the WWW to investigate the current status of the EaStMAN network.

12.6   Figure 12.9 shows an ATM link. Determine the number of possible routes from A to D. Note that in determining the number of routes it should not be possible to return to a node already visited.

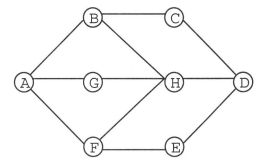

Figure 12.9   Possible routes

12.7   Figure 12.10 shows an ATM link with three possible routes between nodes A and E. The diagram shows the probability of a successful transmission over each segment. Determine the probabilities of a successful transmission over the three routes and thus determine the best route.

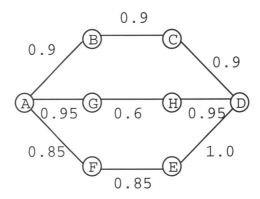

Figure 12.10   Routing probabilities

13

Error control

13.1 INTRODUCTION

Error bits are added to data to either correct or detect transmission errors. Normally, the more bits that are added the better the detection or correction. Error detection allows the receiver to determine if there has been a transmission error. It cannot rebuild the correct data and must either request a retransmission or discard the data. With error correction the receiver detects an error and tries to correct as many error bits as possible. Again, the more error coding bits used the more bits can be corrected. An error correction code is normally used when the receiver cannot request a retransmission.

13.2 PARITY

Parity bits are added to the transmitted data to make the number of 1's sent either even (even parity) or odd (odd parity). It is a simple method of error detection and requires only exclusive-OR (XOR) gates to generate the parity bit. This output can be easily added to the data using a shift register.

Parity bits can only detect an odd number of errors, that is, 1, 3, 5, etc. If an even number of bits are in error then the parity bit will be correct and no error will be detected. This type of coding is normally not used on its own or where there is the possibility of several bits being in error.

13.3 CYCLIC REDUNDANCY CHECKING (CRC)

CRC is one of the most reliable error detection schemes and can detect up to 95.5% of all errors. The most commonly used code is the CRC-16 standard code which is defined by the CCITT.

The basic idea of a CRC is that the transmitter and receiver both agree that the numerical value sent by the transmitter will always be divisible by

9. If the receiver gets a value which is not divisible by 9 then it knows that there has been an error. For example, if a value of 32 is to be transmitted then this value could be changed to 320 so that the transmitter can add to the least significant digit to make it divisible by 9. In this case the transmitter would add 4, making 324. If this transmitted value was to be corrupted in transmission then there would only be a 10% chance that an error would not be detected.

In the CRC-16, the error correction code is 16 bits long and is the remainder of the data message polynomial G(x) divided by the generator polynomial P(x) ($x^{16} + x^{12} + x^5 + 1$, that is, 10001000000100001). The quotient is discarded and the remainder is truncated to 16 bits. This is then appended to the message as the coded word.

The division does not use standard arithmetic division process. Instead of the subtraction operation an exclusive-OR operation is used. This is a great advantage as the CRC only requires a shift register and a few EX-OR gates to perform the division.

The receiver and the transmitter both use the same generating function P(x). If there are no transmission errors then the remainder will be zero.

The method used is as follows:

1. let P(x) be the generator polynomial and M(x) the message polynomial;
2. let n be the number of bits in P(x);
3. append n zero bits onto the right-hand side of the message so that it contains m+n bits;
4. using modulo-2 division, divide the modified bit pattern by P(x). Modulo-2 arithmetic involves exclusive-OR operations, i.e. 0-1=1, 1-1=0, 1-0=1 and 0-0=0;
5. the final remainder is added to the modified bit pattern.

Example

For a 7-bit data code 1001100 determine the encoded bit pattern using a CRC generating polynomial of $P(x) = x^3 + x^2 + x^0$. Show that the receiver will not detect an error if there are no bits in error.

Answer

$P(x) = x^3 + x^2 + x^0$ (1101)

$G(x) = x^6 + x^3 + x^2$ (100100)

Multiply by the number of bits in CRC polynomial.

$$x^3(x^6 + x^3 + x^2)$$
$$x^9 + x^6 + x^5 = 1001100000$$

```
            SENDER                              RECEIVER

                1111101                             1111101
     1101 ⌐1001100000              1101 ⌐1001100001
            1101                                    1101
            1001                                    1001
            1101                                    1101
             1000                                    1000
             1101                                    1101
             1010                                    1010
             1101                                    1101
              1110                                    1110
              1101                                    1101
               1100                                    1101
               1101                                    1101
                001                                     000
   CRC is 001      ↗             No error      ↗
```

Message sent: 1001100 001

The CRC-CCITT is a standard polynomial for data communications systems and can detect:

- all single and double bit errors;
- all errors with an odd number of bits;
- all burst errors of length 16 or less;
- 99.997 % of 17-bit error bursts; and
- 99.998 % of 18-bit and longer bursts.

13.4 LONGITUDINAL / VERTICAL REDUNDANCY CHECKS (LRC/VRC)

RS-232 uses vertical redundancy checking (VRC) when it adds a parity bit to the transmitted character. Longitudinal (or horizontal) redundancy checking (LRC) adds a parity bit for all bits in the message at the same bit position. Vertical coding operates on a single character and is known as character error coding. Horizontal checks operate on groups of characters and is described as message coding. LRC always uses even parity and the parity bit for the LRC character has the parity of the VRC code.

Example

A communications channel uses ASCII character coding and LRC/VRC bits with added to each word sent. Encode the word "Freddy" and, using odd parity for the VRC and even parity for the LRC, determine the LRC character sent.

Answer

	F	r	e	d	d	y	LRC
bo	0	0	1	0	0	1	**0**
b1	1	1	0	0	0	0	**0**
b2	1	0	1	1	1	0	**0**
b3	0	0	0	0	0	1	**1**
b4	0	1	0	0	0	1	**0**
b5	0	1	1	1	1	1	**1**
b6	1	1	1	1	1	1	**0**
VRC	0	1	1	0	0	0	**1**

The character sent for LRC is thus `10101000` or a ')'. The message sent is 'F', 'r', 'e', 'd', 'd', 'y' and '('.

Without VRC checking, LRC checking detects most errors but does not detect errors where an even number of characters has an error in the same bit position. For example in the previous example if bit 2 of the 'F' and 'r' was in error then LRC is still valid.

This problem is overcome if LRC and VRC are used together. With VRC/LRC the only time an error goes undetected is when an even number of bits, in an even number of characters, in same bit positions of each character are in error. This is of course very unlikely.

On systems where only single bit errors occur then the LRC/VRC method can be used to detect and correct the single bit error. For systems where more than one error can occur it is not possible to locate the bits in error and the receiver thus prompts the transmitter to re-transmit the message.

13.5 HAMMING CODE

Forward error correction (FEC) schemes detect and correct bit errors. A popular FEC code is Hamming code. The error correction bits are known as Hamming bits, the number that require to be added to a character is determined by the expression:

$$2^n \geq m + n + 1$$

where :   m is number of bits in the data, and
n is number of Hamming bits.

Hamming bits insert into the message character in a desired way. Typically, they are added in positions of powers of 2, i.e. the 1st, 2nd, 4th, 8th, 16th, and so on, bit positions. For example to code the character 011001 then, starting from the right-hand side, the Hamming bits would be inserted into the 1st, 2nd, 4th and 8th bit position.

The character is   011001
The Hamming bits are   HHHH
Thus format of message will be   01H100H1HH

10	9	8	7	6	5	4	3	2	1
0	1	H	1	0	0	H	1	H	H

Next each position where there is a 1 is represented as a binary value. Then each position value is exclusive-OR'ed with each other. The result is the Hamming code. In this example:

Position	Code
9	1001
7	0111
3	0011
EX-OR	1101

The Hamming code error bits are thus `1101` and the message transmitted will be `0111001101`.

10	9	8	7	6	5	4	3	2	1
0	1	1	1	0	0	1	1	0	1

At the receiver all bit positions where there is a 1 are exclusived-OR'ed. The result gives either the bit position error or no error. If the answer is zero there were no single bit errors, or it gives the bit error position. For example, if we have no transmission errors the result will give zero.

Position	Code
Hamming	1101
9	1001
7	0111
3	0011
EX-OR	0000

If an error has occur in bit 4 then the result is 4.

Position	Code
Hamming	1101
9	1001
7	0111
4	0100
3	0011
EX-OR	0100

13.6 TUTORIAL

13.1   Decode the following asynchronous message. The encoding used is 1 start bit, 1 stop bit, 7-bit ASCII and odd parity.

```
1111110011000101100100111111111111110101000
11111000100110111111111
```

13.2   Determine the number of Hamming bits required to code a single 7-bit ASCII character.

13.3 Determine the Hamming bits for the following 7-bit ASCII characters. Insert the Hamming bits into every other location starting from the left hand side:

(i) NULL
(ii) 'f'
(iii) '9'
(iv) DEL

13.4 For a 7-bit code show how Hamming can detect and correct one error for the following transmitted codes.

(i) 0000000
(ii) 0101010
(iii) 1111111

13.5 Using Hamming code determine the bits sent for the following message (assume 7-bit ASCII coding):

Hamming code

13.6 Determine the message and errors in the following code. Coding used is ASCII with even parity:

```
01001000   01100101   01101100   01101100
01101111   00101111   01010111   01101111
01110010   01101100   01100100   00101110
```

13.7 Determine the errors in the following Hamming error coded data and determine the messageNote 7-bit ASCII coding with Hamming bits in their optimum position:

```
11000110010, 11110011001,
11000101100, 11000101011
```

13.8 Explain why, for single bit errors, that the VRC and LRC, when used together can be used to correct errors.

13.9   Determine the LRC for the word **SCOOBY**, use ASCII coding. Use odd parity for LRC and VRC

13.10   Using CRC coding with a message of `1101011011` and a generator of `10011` determine the transmitted message

13.11   Determine the encoded message for the following 8-bit data codes using the CRC generating polynomial of:

$$P(x) = x^4 + x^3 + x^0$$

(i)   `11001100`
(ii)   `01011111`
(iii)   `10111011`
(iv)   `10001111`

Prove that there will be no error at the receiver.

13.12   Using the polynomial given in Q13.10 determine if the following encoded values have any errors:

(i)   `1111111100110`
(ii)   `1011000011101`
(iii)   `1101110101001`

14

X-Windows

14.1 INTRODUCTION

An operating system allows application programs to access the hardware of computer. A user interface allows a user to access application programs in an easy-to-use way. In a text-based system the user enters text commands which are then interpreted by the operating system. Typical commands are to run application programs, copy files and so on. With a graphical interface, applications programs are represented by graphical objects (icons) and options are selected using menus. Rather than entering commands from the keyboard a mouse pointer is normally used to select objects. This usage of Windows, Icons, Menus and Pointers (WIMPs) is typically known as a Graphical User Interface (GUI).

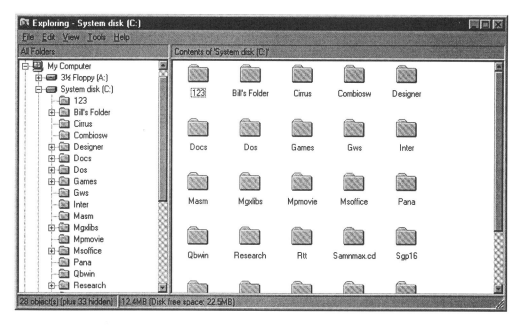

Figure 14.1 Microsoft Windows 95 sample window

The two main graphical user interfaces are Microsoft Windows and X-

Windows. Newer versions of these implement layers 5, 6 and 7 of the OSI model and use standard networking technologies, such as Ethernet and networking protocols like TCP/IP. Both are becoming de facto standards for computer systems. Microsoft Windows 95 is used on PCs and can use TCP/IP to connect to other networks. A typical Windows 95 screen is shown in Figure 14.1.

X-Windows runs on most computer systems but is typically used on Unix workstations. It is a portable user interface and can be used to run programs remotely over a network. Massachusetts Institute of Technology (MIT) developed it and it has become a de facto standard because of its manufacturer independence, its portability, its versatility and its ability to operate transparently across most network technologies.

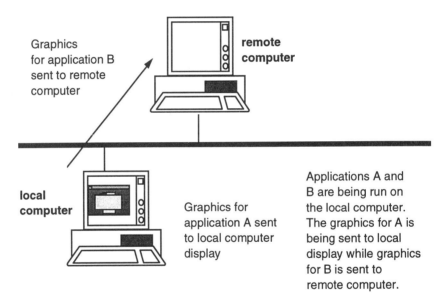

Figure 14.2   X programs can display to remote and local machines

The main features of X-Windows are:

- that it is network transparent. The output from a program can either be sent to the local graphics screen or to a remote node on the network. Application programs can output simultaneously to displays on the network, as illustrated in Figure 14.2. The communication mechanism used is machine-independent and operating system independent.
- that many different styles of user interface can be supported. The man-

agement of the user interface, such as the placing, sizing and stacking of windows is not embedded in the system, but is controlled by an application program which can easily be changed.

- that since X isn't embedded into an operating system it can be easily ported to a wide range of computer systems;
- that calls are made from application programs to the X-windows libraries which control WIMPs. The application program thus does not have to create any of these functions.

14.2 FUNDAMENTALS OF X

There are three main parts of X software:

- a 'server' to control the physical display and input devices;
- 'client' programs which request the server to perform particular operations on specified windows;
- a 'communications channel' through which the client programs and the server talk to each other.

This relationship is shown in Figure 14.3.

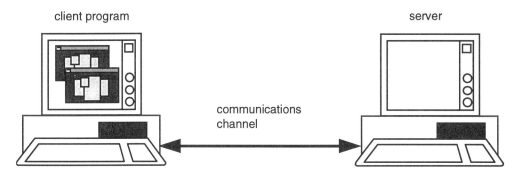

client program   server

communications
channel

Figure 14.3   Fundamentals components of X

14.2.1 Server

The server is the software which creates windows and draw images and text within them. This is done in response to the client programs.

14.2.2 Client

An application program makes use of the system's window facilities. Application programs in X are called 'clients' as they are customers of the server and ask the server to perform task on their behalf. For example a client may request the server to 'display the text Input a value' in window USER1 or to 'draw a rectangle in window TEMP'. This obviously reduces the burden to processing graphics on the client but increasing the processing on the server.

14.2.3 Communications channel

Clients send requests to the server via the communications channel and vice versa. X-Windows is transparent to the network technology and it supports different types of communications channel in the basic X-Windows library. All dependence of the types of communication is isolated to this library, and all communication between client and server is via this library (known as Xlib in the standard X implementation) as illustrated in Figure 14.4.

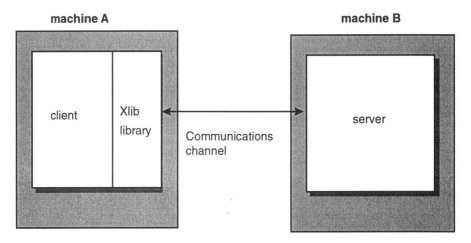

Figure 14.4   Function of the Xlib library

There are two main modes of communications between the client and server, these are:

1. when the server and client are running on the same computer. Here they

can communicate using any method of inter-process communications (ICP) available on the machine. When running in this mode, X is effectively operating like many conventional window systems.

2. when the client is running on one machine, but the display (and its server) is on another. The client and server communicate across the network using a mutually agreed protocol, such as TCP/IP.

X allows programs to communicate user information which is network transparent. This feature is useful in building multi-purpose networks of co-operating machines.

As the server and client are completely separate then a computer can run the application programs and that the client is required to do is to provide the keyboard, pointing device and graphics screen. This has led to a new type of display called an X terminal. An X terminal is simply a stripped-down computer which is dedicated to running the X server and nothing else. It has a keyboard, mouse and screen and some way of communicating across the network. It does not have its own file-system and cannot support general purpose programs. Consequently, these programs have to be run elsewhere on the network.

14.3 NETWORK ASPECTS OF X

The main advantages of the client/server relationship are:

- that a client can use a powerful remote server. This server could be a supercomputer, have a special processor, enhanced floating-point accelerator, and so on. An example of several X terminals and a computer using powerful workstations is illustrated in Figure 14.5;
- that if the server is a file server providing most of the disk resources to the local network then the disk network traffic is reduced as applications are run remotely on the server particular applications which are highly disk-intensive such as a large application programs. In this way only the results from a program need to be reported over the network and intermediate data is not required to be sent over the network.
- the remote server may have special software facilities available on it alone. With the growth of workstations, it is increasingly common to have some software licensed to only a few machines on the network

and several clients run the software from these licensed machines. This may reduce the licence fee as only the servers may need to be paid for.

- it allows an enhanced user interface for remote logging into networked machines;
- when a new computer is added which can run X clients, it can be immediately used by any device running X.
- if a new display is added it can immediately make use of all the existing X client applications on any machine.
- there is a requirement to output to several displays.

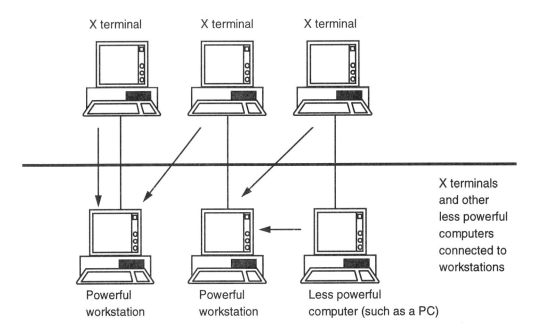

Figure 14.5 Client and server on separate machines

The main disadvantage of X-Windows is that application programs become dependent on servers. If a server becomes overburdened or develops a fault then it can seriously affect the performance of application programs. This problem is similar to network problems in star network where the network performance is dependent upon the central server.

The more programs that run on the server then the more main memory it needs to cope with the loading. If it does not have enough main memory then it can use a local disk to compensate. This will slow the performance of application programs.

An X-based network must thus be planned so as not to burden the server nodes. Typically, only two X terminals connect to a server. A well-planned network can normally cope with server faults by having several computers which can act as servers to X terminals, or remote computers.

14.4 HISTORY OF X

In 1984, MIT started the development of X. Their main objective was to create a good windows system for Unix machines. Many versions evolved from this and by 1985 it was decided that X would be available to anyone who wanted for a nominal cost.

   In 1985, version 10 was distributed to organizations outside MIT and by 1986 DEC produced the first commercial X product. During that year it was clear that version 10 could not evolve to satisfy all the requirements that were being asked for it. For this purpose MIT and DEC undertook a complete redesign of the protocol. The result was X, version 11, this has since become known as X.11.

14.5 X SYSTEM PROGRAMS

Any program can use the X-Windows libraries. Several standard programs exist which comprise the basic system:

- X is the display server. This software controls the keyboard, mouse and screen. It is the heart of X and creates and destroys windows. It writes text and draws objects within windows at the request of other 'client' programs. X is normally started by the command xstart or xinit.
- xterm is the X terminal emulator. Many programs were not written specifically for X-Windows and must be run within a text window. The xterm program converts the text output to the graphics display.
- xhost controls which other hosts on the network are allowed access to display screen.
- xkill kills unwanted applications.
- xwd captures an image from the screen.
- xpr prints a previously captured screen to the printer.
- xmag magnifies a selected portion of the screen.
- xclock to display an analogue or digital clock.
- xcalc to be used a calculator.

- xload to displays the system loading of the machine.

14.6 TUTORIAL

14.1 Discuss the advantages and disadvantages of a network using X-Windows.

14.2 Why is manufacturer independence an advantage with the user interface ?

14.3 A sample network is shown in Figure 14.6. The upper network contains a PC file server and 30 PCs. This network also contains a number of X-terminals which are accessing servers on the lower network. The lower network contains two server computers and a number of X-terminals. Discuss the problems that could occur with this set-up.

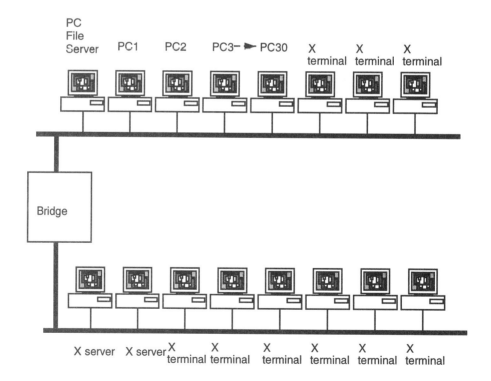

Figure 14.6 Tutorial Q14.3

14.4 For a known network, determine the operating system and user interface used on the systems on the network. Also determine what typical of network protocol it uses.

15

Analysis of digital pulses

15.1 INTRODUCTION

Information is transmitted as an energy from a source to a destination. This energy can take the form of light waves, radio waves or even sound waves. Any electronic signal can be analyzed either in the time domain or in the frequency domain. All electrical signals, no matter their shape, can be represented by a series of sine or cosine waves.

The standard form of a single frequency signal is:

$$V(t)=V \sin(2\pi ft+\theta)$$

where $v(t)$ is the time varying voltage (V), V is the peak voltage (V), f the frequency (Hz) of the signal and θ its phase (°)

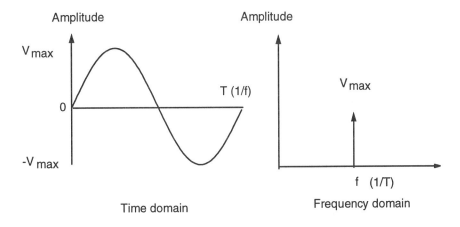

Figure 15.1 Representation of signal in frequency and time domains

A signal can be represented in the time domain as a varying voltage against time or in the frequency domain as voltage amplitudes against frequency. Figure 15.1 shows how a sine wave is represented in the time domain and the frequency domain. The signal shown has a period T, the frequency of the signal will be 1/T Hz. This is shown in the frequency domain as a single vertical arrow at that frequency. The amplitude of the arrow represents the amplitude of the signal.

15.2 REPETITIVE SIGNALS

A repetitive signal is one that repeats after a given time. It can be shown that a repetitive signal is made up of a series of sine and/or cosine waves, called the Fourier series. It can be described by:

$$f(t) = A_0 + A_1 \cos\omega_1 t + A_2 \cos 2\omega_1 t + ... + A_N \cos N\omega_1 t$$
$$+ B_1 \sin\omega_1 t + B_2 \sin 2\omega_1 t + ... + B_N \sin N\omega_1 t$$

where ω_1 is the fundamental angular frequency ($=2\pi f_1$).

This equation shows that the waveform comprises of an average value (A_0), a series of cosine functions in which each successive term has a frequency that is an integer multiple of the frequency of the first cosine (or sine) in the series. The A and B components can either be found using tables or by using the mathematical formula:

$$A_0 = \frac{1}{T} \int f(t) dt$$

$$A_N = \frac{2}{T} \int f(t) \cdot \sin(N\omega_1 t) dt$$

$$B_N = \frac{2}{T} \int f(t) \cdot \sin(N\omega_1 t) dt$$

Any periodic waveform has an average, or DC, component and a series of harmonically related sine and cosine waves. A harmonic is an integral multiple of the fundamental frequency. The first harmonic is the fundamental frequency, the second is twice the frequency of the fundamental, the third is three times the multiple, and so on. The fundamental frequency is the lowest frequency in the signal and is thus equal to the inverse of the repetition time. Thus a periodic waveform can be represented by:

f(t) = DC + fundamental + 2nd harmonic + 3rd harmonic + ... + nth harmonic

An example of a repetitive wave is given in Figure 15.2. It contains a fundamental frequency of amplitude 1 V, a third harmonic amplitude of 0.3 V and fifth harmonic amplitude of 0.2 V. The equation for this wave is:

$$f(t) = \sin(\omega_1 t) + 0.3 \sin(3\omega_1 t) + 0.2 \sin(5\omega_1 t)$$

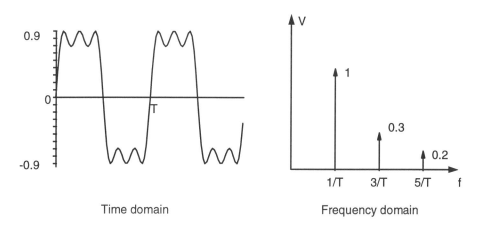

Time domain   Frequency domain

Figure 15.2   Time and frequency domain representation of a repetitive signal

15.3 WAVE SYMMETRY

If a periodic signal is symmetrical about either the vertical or horizontal axis then either the cosine terms or the sine terms become zero.

15.3.1 Even symmetry

When a periodic signal is symmetrical about the vertical axis then it is an even function and the B coefficients in the Fourier equation become zero. Thus the waveform contains only cosine components and a DC level. An example of this type of waveform is given in Figure 15.3.

With this function f(t) = f(−t), thus the resulting equation will be:

$$f(t) = A_0 + A_1 \cos\omega_1 t + A_2 \cos 2\omega_1 t + \ldots + A_N \cos N\omega_1 t$$

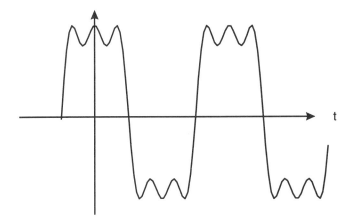

Figure 15.3 Even symmetry

15.3.2 Odd symmetry

When a periodic signal is symmetrical about the line midway between the vertical and horizontal axis it is an odd function and the A coefficients in the Fourier equation are then zero. Thus the waveform will contain only sine components, with no DC offset. An example of this type of waveform is given in Figure 15.4.

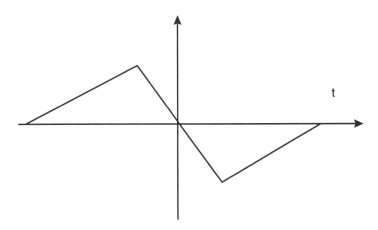

Figure 15.4 Odd symmetry

With this function f(t) = –f(–t), thus the resulting equation will be:

$$f(t)=B_1 \sin\omega_1 t+B_2 \sin 2\omega_1 t+...+B_N \sin N\omega_1 t$$

15.3.3 Half-wave symmetry

When the second half cycle of periodic signal is the same as the first half, but is the inverse, then it has half-wave symmetry. The even harmonics in this wave become zero and the waveform will only contain odd harmonics (1st, 3rd, 5th, .., and so on).

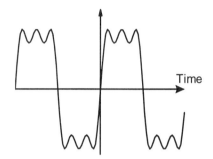

Figure 15.5   Half-wave symmetry

15.4 FOURIER SERIES OF A REPETITIVE RECTANGULAR WAVEFORM

The signal shape of most interest in data communications is the repetitive rectangular pulse, as shown in Figure 15.6. It is defined by its amplitude and its duty cycle, which is the ratio of the active time of the pulse (τ) to the period of the waveform (T). The duty cycle is thus given by:

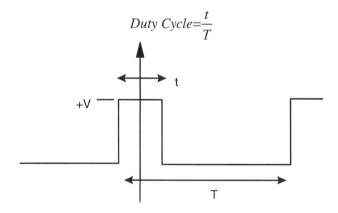

Figure 15.6   Repetitive pulse waveform

The time-based repetitive pulse waveform is given by:

$$v(t) = \frac{V\tau}{T} + \sum_{n=1}^{n=\infty} V_n \cos(n2\pi f_1 t)$$

the amplitudes of the harmonics is given by:

$$V_n = \frac{2V\tau}{T} \cdot \frac{\sin Nx}{x}$$

where

$$x = \frac{\pi\tau}{T}$$

V_1 is the amplitude of the fundamental, V_2 is the amplitude of the second harmonic, etc. The frequencies contained in the signal will be:

$$f_1 = \frac{1}{T} \text{ Hz}, f_2 = \frac{2}{T} \text{ Hz}, f_3 = \frac{3}{T} \text{ Hz, etc.}$$

The DC component of the signal is thus:

$$V.\frac{\tau}{T}$$

The RMS voltage of a repetitive signal with peak voltage harmonics V_1.. V_n and DC component V_0 is given by the formula:

$$V_{rms} = \sqrt{V_0^2 + \frac{V_1^2}{2} + \frac{V_2^2}{2} + \ldots \ldots \frac{V_n^2}{2}}$$

where V_0 is the DC voltage, V_1 the peak amplitude of first harmonic, and so on. It can be seen that the amplitudes of the harmonics varies as the sin(x)/x function. A typical sin(x)/x function is shown in Figure 15.7.

Figure 15.8 gives an example of a repetitive pulse train with a duty cycle of 0.2 and a pulse amplitude of 1 V.

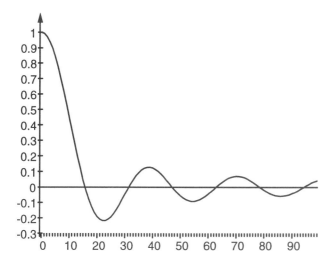

Figure 15.7   Sin(x)/x function

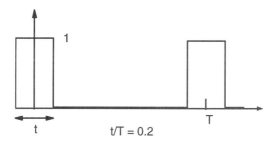

Figure 15.8   Pulse train with a duty cycle of 0.2

The corresponding Fourier series is given by:

$$v(t) = \frac{Vt}{T} + \sum_{N=1}^{\infty} \left[\frac{2Vt}{T} \cdot \frac{\sin(N\pi \, {}^{t}\!/_{T})}{N\pi \, {}^{t}\!/_{T}} \right] \cos(N\omega t)$$

Figure 15.9 shows the amplitudes of the frequency harmonics.

15.5 EXAMPLES

Repetitive pulses of 5 V amplitude, pulse width of 5 μs and repetition time of 25 μs is applied to a communications channel which can be modelled as an ideal low-pass filter with a pass band up to 140 kHz. Figure 15.10 shows the pulse train.

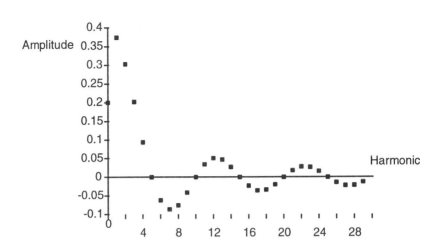

Figure 15.9   Frequency spectrum (for duty cycle of 0.2)

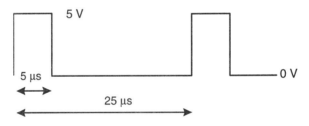

Figure 15.10   Repetitive pulse train

Determine:

(i)   DC voltage offset of the input signal;
(ii)   first five harmonic frequencies of the input signal;
(iii)   amplitude of the first five harmonics in the input signal.

Also, sketch the time domain response, over a period of 25 μs, of the output signal.

ANSWER
The time response will be:

$$v(t)=\frac{Vt}{T}+\sum_{N=1}^{\infty}\left[\frac{2Vt}{T}\cdot\frac{\sin(N\pi\,{}^{t}\!/_{T})}{N\pi\,{}^{t}\!/_{T}}\right]\cos(N\omega_1 t)$$

(i) DC offset:

$$V_{DC} = V_{pk} \frac{t}{T} = 5 \cdot \frac{5}{25} = 1 \ V$$

(ii) First five frequencies:

$$f_1 = \frac{1}{T} = \frac{1}{25 \times 10^{-6}} = 40 \ kHz$$
$$f_2 = 80 \ kHz$$
$$f_3 = 120 \ kHz$$
$$f_4 = 160 \ kHz$$
$$f_5 = 200 \ kHz$$

(iii) Amplitude of first five harmonics:

$$V_N = \frac{2Vt}{T} \cdot \frac{\sin(N\pi \, {}^t\!/_T)}{N\pi \, {}^t\!/_T}$$

Thus:

$$V_N = \frac{2 \times 5 \times 5}{25} \cdot \frac{\sin(0.2N\pi)}{0.2N\pi}$$
$$= \frac{3.18}{N} \cdot \sin(0.63N) \quad V$$

Thus:

N	f (kHz)	V amplitude (Volts)
1	40	1.87
2	80	1.51
3	120	1.01
4	160	0.47
5	200	0

$$v_i(t) = 1 + 1.87 \sin(\omega_1 t) + 1.51 \sin(2\omega_1 t) + 1.01 \sin(3\omega_1 t)$$
$$+ 0.47 \sin(4\omega_1 t) + \ldots\ldots$$

assuming filter blocks above 140 kHz, then the output will be:

$$v_o(t) = 1 + 1.87 \sin(\omega_1 t) + 1.51 \sin(2\omega_1 t) + 1.01 \sin(3\omega_1 t) \quad V$$

This gives the following table:

ωt (°)	V_0	V_1	V_2	V_3	Σ
	1	1.77 cosωt	1.51 cos2ωt	1.01 cos3ωt	
45	1	1.32	0	−0.71	1.61
90	1	0	−1.51	0	−0.51
135	1	−1.32	0	0.71	0.39
180	1	−1.87	1.51	−1.01	−0.37
225	1	−1.32	0	0.71	0.39
270	1	0	−1.51	0	−0.51
315	1	1.32	0	−0.71	1.61
0,360	1	1.87	1.51	1.01	5.39

The pulse output time response can now be plotted for one cycle. Figure 15.11 shows a rough sketch of the output pulse. The shape of the output would be much smoother if more time points were taken.

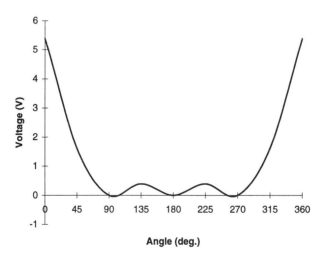

Figure 15.11   Pulse output

The frequency domain of the output has a DC value of 1 V, a fundamental frequency of 40 kHz, amplitude 1.87 V; a second harmonic at 80 kHz, amplitude 1.51 V; a third harmonic at 120 kHz, amplitude 1.01 V; a forth harmonic at 160 kHz, amplitude 0.47 V and there is no fifth harmonic. A diagram of this is given in Figure 15.12.

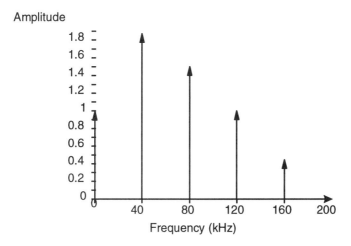

Figure 15.12   Frequency response of output

15.6 PROGRAM TO DETERMINE HARMONICS

Program 15.1 is a C program which determines the harmonics for a square wave or a rectangular pulse waveform.

▤ **Program 15.1**
```c
/*   Program to determine the harmonics of      */
/*   repetitive rectangular or square pulses    */

#define   PI      3.14159
#define   TRUE    1
#include <math.h>
#include <stdio.h>

void    Square_wave(int harm,float Vmax);
void    Pulse_wave(int harm,float Vmax);
int     SelectOption(void);

int     main(void)
{
int     select,harmonics;
float   Vmax;

/* infinite loop until break */

   while (TRUE)
   [
     select=SelectOption();
     if (select==3) break;

     puts("Enter number of harmonics required");
```

```
        scanf("%d",&harmonics);

        puts("Enter max voltage ");
        scanf("%f",&Vmax);

        switch (select)
        {
           case 1: Square_wave(harmonics,Vmax);   break;
           case 2: Pulse_wave(harmonics,Vmax);    break;
        }
    }
    return(0);
}

void    Square_wave(int harm,float Vmax)
{
float   v;
int     i;

    for (i=1;i<=harm;i+=2)
    {
       v=(4*Vmax/i/PI);
       printf("Harmonic %d Amplitude %.3f\n",i,v);
    }
}

void    Pulse_wave(int harm,float Vmax)
{
float   Duty,v;
int     i;

    puts("Enter duty cycle (0->1)");
    scanf("%f",&Duty);

    v=(Vmax*Duty);
    printf("DC %.3f\n",v);
    for (i=1;i<=harm;i++)
    {
       v=(2*Vmax*Duty) * sin(i*PI*Duty) / (i*PI*Duty);
       printf("Harmonic %d Amplitude %.3f\n",i,v);
    }
}

int     SelectOption(void)
{
int     Select;

    do
    {
       puts("Do you wish");
       puts("1- Square wave");
       puts("2- Pulse train");
       puts("3- Exit");
       scanf("%d",&Select);

    } while ( (Select<0) || (Select>3) );

    return(Select);
}
```

Test run 15.1 shows a sample run.

🖳 **Test run 15.1**
```
Do you wish
1- Square wave
2- Pulse train
3- Exit
1
Enter number of harmonics required
10
Enter max voltage
4
Harmonic 1 Amplitude 1.274
Harmonic 3 Amplitude 0.425
Harmonic 5 Amplitude 0.255
Harmonic 7 Amplitude 0.182
Harmonic 9 Amplitude 0.142
************************

Do you wish
1- Square wave
2- Pulse train
3- Exit
2
Enter number of harmonics required
10
Enter max voltage
1
Enter duty cycle (0->1)
0.2
DC 0.200000
Harmonic 1 Amplitude 0.374
Harmonic 2 Amplitude 0.303
Harmonic 3 Amplitude 0.202
Harmonic 4 Amplitude 0.094
Harmonic 5 Amplitude 0.000
Harmonic 6 Amplitude -0.062
Harmonic 7 Amplitude -0.086
Harmonic 8 Amplitude -0.076
Harmonic 9 Amplitude -0.042
Harmonic 10 Amplitude -0.000
*************************
Do you wish
1- Square wave
2- Pulse train
3- Exit
3
```

15.7 TUTORIAL

15.1   For repetitive waveforms in Figure 15.13 determine the DC offset.

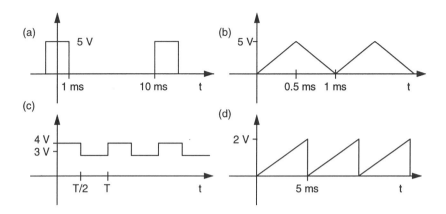

Figure 15.13 Question 15.1

15.2 For repetitive waveforms in Figure 15.14 determine they have a DC offset and if they are made up of sines or cosines.

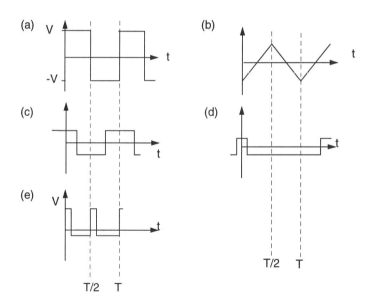

Figure 15.14 Question 15.2

15.3 For the repetitive waveforms in Figure 15.15 determine the following:

(i) the DC offset;
(ii) the first 10 harmonics amplitudes;

(iii) a sketch of the frequency spectrum;
(iv) the time response using the DC offset using he first five harmonics.

15.4 Determine the shape of frequency spectrum for the following duty cycles.

(i) Duty cycle = 0.25
(ii) Duty cycle = 0.125
(iii) Duty cycle = 0.03125

Display the first 13 harmonics and discuss the outline. Determine which pulse would be most affected when passed through a low-pass filter.

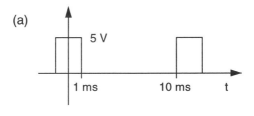

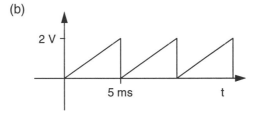

Figure 15.15 Question 15.3

15.5 A square wave, with a repetition time of 0.2 ms, no DC offset and a peak to peak voltage of 2 V, is applied to a low-pass filter which blocks frequencies above 6 kHz. Determine the following:

(i) DC voltage output;
(ii) peak voltage output;
(iii) RMS voltage output;
(iv) power output assuming a 1 Ω load.

16

Transmission lines

16.1 INTRODUCTION

Digital pulses are affected by transmission systems in the following ways:

- they are attenuated along the line;
- the transmission line acts as a low-pass filter, blocking high frequencies;
- different frequencies within pulses travel at different rates causing phase distortion of the pulse;
- spreading of the pulses, causing them it to interfere with other pulses;
- mismatches on the line cause reflections (and thus 'ghost pulses').

16.2 EQUIVALENT CIRCUIT

A transmission line transmits electrical signals from a source to a receiver. It can be a coaxial cable, a twisted-pair cable, a waveguide, etc. From a circuit point of view, the conductors of a transmission line contain series resistance and inductance and the insulation between conductors has a shunt conductance and capacitance.

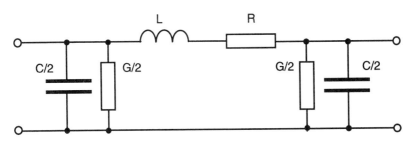

Figure 16.1 Equivalent circuit of a transmission line

If a given length of this transmission line were divided into more and more sections, the ultimate case would be an infinitesimal section of the basic elements, resistance R, conductance G, inductance L and capacitance C. The circuit that results is given in Figure 16.1. Parameters R, L, G and C are known as the primary line constants and are defined as:

- series resistance R Ω.metre^{-1}
- series inductance L   H.metre^{-1}
- shunt conductance G   S.metre^{-1} (Siemens.metre^{-1})
- shunt capacitance C   F.metre^{-1}

The characteristic impedance (Z_0) is the ratio of the voltage to the current for each wave propagated along a transmission line, and is given by:

$$Z_0 = \frac{V}{I} \; \Omega$$

From this it can be shown that:

$$Z_0 = \sqrt{\frac{R + j\omega L}{G + j\omega C}} \; \Omega$$

The characteristic impedance can either be measured by using an infinite length of transmission line and measuring the input impedance or by suitably terminating the line, as shown in Figure 16.2.

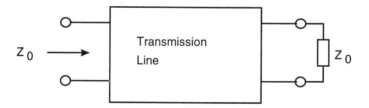

Figure 16.2   Transmission line terminated with Z_0

The magnitude of the characteristic impedance will thus be given by:

$$Z_0 = \sqrt{\frac{R^2 + (2\pi f L)^2}{G^2 + (2\pi f C)^2}} \; \Omega$$

The characteristic impedance can be approximated for certain conditions:

(i)   when the frequency is large, then

$$\omega L \gg R$$
$$\omega C \gg G$$

thus

$$Z_o = \sqrt{\frac{L}{C}} \ \Omega$$

(ii)   on a lossless transmission line then R = 0 and G = 0, thus:

$$Z_o = \sqrt{\frac{L}{C}} \ \Omega$$

16.2.1 Program to determine characteristic impedance

Program 16.1 determines the magnitude of the characteristic impedance for a transmission line with entered primary line constants.

Program 16.1

```
/*    Program to determine impedance of a transmission      */
/*    line impedance of TL                                  */
#include <stdio.h>
#include <math.h>

#define MILLI         1e-3
#define MICRO         1e-6
#define INFINITYFLAG  -1
#define PI            3.14159

float   calc_Zo(float r,float l,float g,float c,float f);
float   calc_mag(float x,float y);
float   calc_imp(float f, float val);

int     main(void)
{
float   Zmag,R,L,G,C,f;

    puts("Program to determine impedance of a transmission line");
    printf("Enter R,L(mH),G(mS),C(uF) and freq.>>");
    scanf("%f %f %f %f %f",&R,&L,&G,&C,&f);
    Zmag=calc_Zo(R,L*MILLI,G*MILLI,C*MICRO,f);
    if (Zmag==INFINITYFLAG)
        printf("Magnitude is INFINITY ohms\n");
    else
        printf("Magnitude is %.2f ohms\n",Zmag);
    return(0);
}
```

```
float   calc_Zo(float r,float l,float g,float c,float f)
{
float   value1,value2;

    value1=calc_mag(r,calc_imp(f,l));
    value2=calc_mag(g,calc_imp(f,c));
    /* Beware if dividing by zero */
    if (value2==0)   return(INFINITYFLAG);
    else             return(sqrt(value1/value2));
}

float   calc_mag(float x,float y)
{
    return(sqrt((x*x)+(y*y)));
}

float   calc_imp(float f,  float val)
{
    return(2*PI*f*val);
}
```

Test run 16.1 shows a sample test run.

⌨ **Test run 16.1**

```
Program to determine impedance of a transmission line
Enter R, L (mH), G (mS), C (uF) and freq.>> 0 40 0 7 1000
Magnitude is 75.59 ohms
```

16.2.2 Propagation coefficient, γ

The propagation coefficient γ determines the variation of current, or voltage, with respect to distance x along a transmission line, as shown in Figure 16.3. The current (and voltage) distribution along a matched line is found to vary exponentially with distance, as given next:

$$I_x = I_s e^{-\gamma x}$$
$$V_x = V_s e^{-\gamma x}$$

where I_s is the magnitude of current at x=0 and V_s the magnitude of voltage at x=0.

Like the characteristic impedance, the propagation coefficient also dependent on the primary constants and the frequency of the signal, and is given by:

$$Z_o = \sqrt{(R+j\omega L)(G+j\omega C)}$$

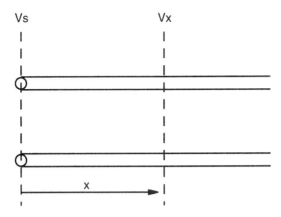

Figure 16.3   Voltage at a distance x

This is a complex quantity and can be written as:

$$\gamma = \alpha + j\beta$$

The attenuation coefficient, α (nepers.metre^{-1}), determines how the voltage or current amplitude varies with distance along the line. The phase shift coefficient, β (radians.metre^{-1}), determines the phase angle of the voltage (or current) variation with distance. Since a phase shift of 2π radians (or 360°) occurs over a distance of one wavelength, λ, then

$$\beta = \frac{2\pi}{\lambda}$$

where λ is the physical wavelength.

16.3 SPEED OF PROPAGATION

The velocity of propagation u is given by:

$$u = f\lambda = \frac{2\pi}{\beta} \quad \text{m.s}^{-1}$$

On a lossless line, or at high frequencies then:

$$\beta = \omega\sqrt{LC}$$

$$u = \frac{f2\pi}{\omega\sqrt{LC}}$$

$$= \frac{1}{\sqrt{LC}} \quad \text{m.s}^{-1}$$

It can be found using field theory and the geometry of normal lines to calculate the inductance and capacitance that:

$$u = \frac{1}{\sqrt{\mu\varepsilon}} = \frac{1}{\sqrt{\mu_o\varepsilon_o\varepsilon_r}} = \frac{c}{\sqrt{\varepsilon_r}} \quad \text{m.s}^{-1}$$

since $\mu_r = 1$ and $c = \dfrac{1}{\sqrt{\mu_o\varepsilon_o}}$. Where c is the velocity of light (m.s^{-1}) and ε_r is the dielectric constant of transmission line.

16.4 TRANSMISSION LINE REFLECTIONS

When a pulse meets a mismatch in an electrical circuit a reflected pulses is bounced back of the mismatch. This 'ghost' pulse is then back along the transmission line. They also cause a loss of signal power.

The characteristic impedance of transmission lines and terminations is the important factor in minimizing reflections and maximizing power transfer. Typically transmission lines, such as coaxial cables, have characteristic impedances of 50 Ω, for TV and video it 75 Ω.

16.4.1 Reflections from resistive terminations

A matched termination is when a transmission line is terminated with a resistance equal to Z_0. The line then behaves as if it has infinite length and there is thus no reflected energy.

If the line is terminated in any resistance other than Z_0, then energy is reflected from the termination. The amplitude of a reflected pulse can be determined by the impedance of the load and the characteristic impedance of the line.

The transmission line in Figure 16.4 is terminated with a resistance which is greater that Z_0. The lattice and surge diagrams show the amplitude of the pulse against time. On the surge diagram pulses are plotted against a horizontal time axis. In the lattice diagram time is plotted on the vertical axis. Both diagrams show the pulse approaching the load. In Figure 16.5 the pulse has reached the load and part of the pulse is reflected. The rest of the pulses energy is transmitted to the load.

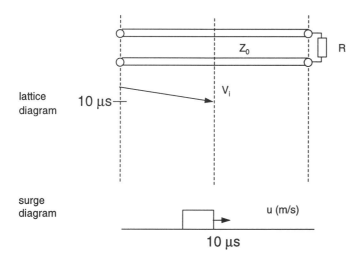

Figure 16.4 Pulse as it approaches the load

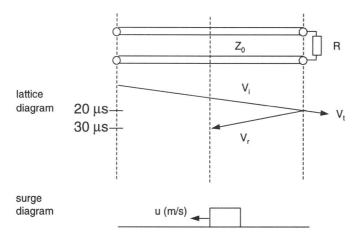

Figure 16.5 Reflected pulse from load

V_i is the incident pulse voltage, V_r the reflected pulse voltage and V_t the transmitted pulse voltage. It can be shown that:

$$V_t = V_i + V_r$$

using this the reflection coefficient (ρ) can be derived:

$$\rho = \frac{V_r}{V_i} = \frac{Z_L - Z_o}{Z_L + Z_o}$$

If the load impedance is equal to the characteristic impedance (Z_0) then no pulse is reflected from the load (that is, $\rho=0$). The transmitted pulse voltage will have the same voltage as the incident pulse.

When the load impedance is larger than the transmission line characteristic impedance then the reflected pulse is positive (that is, $\rho>0$). This pulse travels back along the line towards the source. The resulting transmitted pulse voltage will be larger than the incident pulse as it is the addition of the incident and the reflected pulse voltage.

When the load impedance is less than the transmission line characteristic impedance then the pulse reflected is negative (that is, $\rho<0$) and travels back along the line toward the source. In this case, the transmitted pulse will be less than the incident pulse.

Note that although the transmitted pulse voltage is increased or decreased the electric current also changes. As the voltage increases and the current decreases (or vice versa) it can be proved that there is no increase in electric power. In fact, the reflected pulse gives the loss in transmitted power. It can be shown that the reflection coefficient for current is equal to the negative of the voltage reflection coefficient.

16.4.2 Reflections at junctions between two transmission lines

If two transmission lines are joined, one with a characteristic impedance of Z_{01} and the other of Z_{02} then the reflection coefficient is given by:

$$\rho = \frac{Z_{02} - Z_{01}}{Z_{02} + Z_{01}}$$

For example, if a pulse of 3 V travelling along a cable with $Z_{01} = 50\ \Omega$ meets a cable with characteristic impedance $Z_{02} = 100\ \Omega$, as shown in Figure 16.6, then the reflected and transmitted pulses can be found by:

$$\rho = \frac{100 - 50}{100 + 50} = \frac{1}{3} = \frac{V_r}{V_i}$$

$$V_r = \rho V_i = \frac{1}{3}.3 = 1 \ V$$

$$V_t = V_i + V_r$$
$$\quad = 3 + 1 = 4 \ V$$

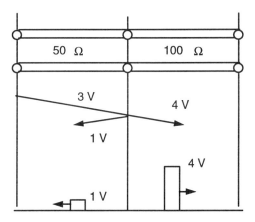

Figure 16.6   Reflection from a junction

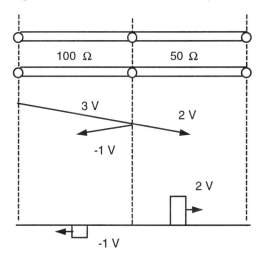

Figure 16.7   Reflection from a junction

If the cables were changed so that Z_{01} is 100 Ω and Z_{02} is 50 Ω, as shown in Figure 16.7, then the reflected and transmitted pulses would be determined by:

$$\rho = \frac{50 - 100}{50 + 100} = -\frac{1}{3}$$

$$V_r = \rho V_i = -\frac{1}{3}.3 = -1V$$

$$V_t = V_i + V_r$$
$$= 3 - 1 = 2V$$

16.4.3 Reflections at junctions with two transmission line in parallel

In many situations two or more transmission lines are connected in parallel to a source transmission line. Figure 16.8 shows the two parallel lines connected to a single source line. If the source line has a characteristic impedance of Z_{01} and the two parallel line have characteristic impedances of Z_{02} and Z_{03}, then the equivalent input impedance at the termination is Z_{02} in parallel with Z_{03}. The equivalent load impedance at the junction between the source line and the parallel lines will be:

$$Z_p = \frac{Z_{02}Z_{03}}{Z_{02} + Z_{03}} \Omega$$

and the reflection coefficient by:

$$\rho = \frac{Z_p - Z_{01}}{Z_p + Z_{01}}$$

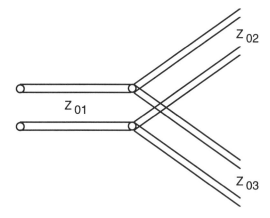

Figure 16.8   Connection to parallel transmission lines

16.5 MATCHING TERMINATIONS

Terminations can be matched by either inserting a series or a parallel resistance. If the characteristic impedance of the source line is higher than the connecting characteristic impedance then a resistance equal to the difference in the impedance is added in series with the connecting line. There will then be no reflections from the load transmission line (although there is a loss of power).

If the characteristic impedance of the source line is lower than the load characteristic impedance then a resistor is inserted in parallel with the junction. To match the junction the equivalent input impedance of the termination should be equal to that of the connecting line as shown in Figure 16.9.

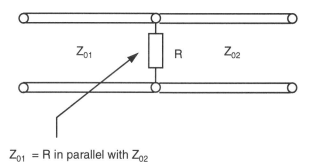

Z_{01} = R in parallel with Z_{02}

Figure 16.9   Matching a termination

If a transmission line with characteristic impedance Z_{02} is the connected to a line with characteristic impedance Z_{01} and the pulse originates from the line with characteristic impedance Z_{01}, then:

if $Z_{01} > Z_{02}$ then series resistor required is $Z_{01} - Z_{02}$ Ω;

if $Z_{01} < Z_{02}$ then parallel resistor required is $\dfrac{Z_{02}Z_{01}}{Z_{02} + Z_{01}}$ Ω.

Example
(a) A uniform transmission line is 100 km long and has the following primary constants per km:

$R = 0$ $\Omega.\text{km}^{-1}$

L = 40   mH.km^{-1}
C = 7   μF.km^{-1}
G = 0   S.km^{-1}

If a digital pulse of 10 V is applied at the input of the line and the receiving end has a 125 Ω equivalent load, calculate:

(i)   speed of propagation of the pulse;
(ii) time for pulse to reach load;
(iii) the characteristic impedance of the line;
(iv) the reflected pulse amplitude;
(v)   the transmitted pulse;
(vi) the value of resistor required to match load.

ANSWER

(i)   speed of propagation (v):

$$v = \frac{1}{\sqrt{LC}}$$

$$= \frac{1}{\sqrt{40 \times 10^{-3} \times 7 \times 10^{-6}}}$$

$$= 1889.8 \ \text{km.s}^{-1}$$

(ii) time taken (t):

$$\text{speed} = \frac{\text{distance}}{\text{time}}$$

$$\text{time} = \frac{\text{distance}}{\text{speed}}$$

$$\text{time taken} = \frac{d}{v} = \frac{100 \times 10^{-3}}{1889.8 \times 10^{3}}$$

$$= 52.9 \ ms$$

(iii) the characteristic impedance (Z_0):

$$Z_0 = \sqrt{\frac{R + j\omega L}{G + j\omega C}}$$

$$Z_0 = \sqrt{\frac{L}{C}}$$

$$Z_0 = \sqrt{\frac{40 \times 10^{-3}}{7 \times 10^{-6}}}$$
$$= 75.6 \ \Omega$$

(iv) the reflected pulse amplitude (V_r):

$$\rho = \frac{Z_L - Z_0}{Z_L + Z_0}$$
$$= \frac{125 - 75.6}{125 + 75.6}$$
$$= 0.25$$
$$V_r = \rho V_i$$
$$= 0.25 \times 10$$
$$= 2.5 \ V$$

(v) the transmitted pulse amplitude (V_t):

$$V_t = V_i + V_r$$
$$= 10 + 2.5$$
$$= 12.5 \ V$$

(vi) resistor to match load (R):

$$\frac{125.R}{125 + R} = 76.5$$
$$125R = 75.6R + 9450$$
$$49.4R = 9450$$
$$R = 191.3 \ \Omega$$

16.6 OPEN AND SHORT CIRCUIT TERMINATIONS

A short circuit has a zero impedance. If this were used as a load then the reflection coefficient is −1. The impedance of an open circuit is infinity, and gives a reflection coefficient of +1. Thus with an open circuit load, the reflected pulse is equal to the incident pulse. For a short circuited load the reflected pulse is the same magnitude, but will be negative. The proofs for the reflection coefficients for an open and short circuit load are given next:

$$\rho_{S/C} = \frac{Z_L - Z_0}{Z_L + Z_0}$$

$$= \frac{0 - Z_0}{0 + Z_0}$$

$$= -1$$

$$\rho_{O/C} = \frac{\infty - Z_0}{\infty + Z_0}$$

$$= \frac{\infty}{\infty}$$

$$= +1$$

A typical technique used to find faults on underground cables is to send a pulse along the line and measure the time taken for a reflection pulse to return. Since the speed of propagation of the pulse is known then the distance to the fault can be found by dividing the speed of propagation by half the time taken (or twice the speed of propagation divided by the time taken). For a short circuit on the line an inverted voltage pulse is returned, else, if there is an open circuit a positive voltage pulse is returned, as illustrated in Figure 16.10.

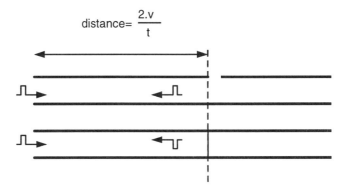

Figure 16.10   Testing for faults on a line

16.7　TUTORIAL

16.1　In the diagram in Figure 16.11, R_S is 50 Ω and Z_L is an open-circuit. A 10 V, 1 µs pulse is applied to the line on the entry to the line. Determine the voltages on the line at the time intervals 1 µs, 3 µs and 5 µs. The time taken for the pulse to travel from the source to the load is 2 µs. Assume that the pulse voltage entering the 50 Ω line is always 10 V.

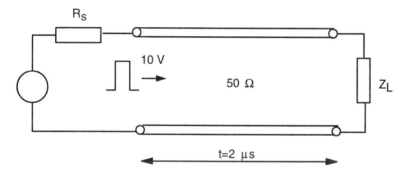

Figure 16.11　Tutorial question

16.2　Repeat problem 16.1, with Z_L as a short circuit.

16.3　Repeat problem 16.1, with R_S as 450 Ω and Z_L a short circuit.

16.4　Repeat problem 16.1, with R_S as 450 Ω and Z_L an open circuit.

16.5　Repeat problem 16.1, with R_S and Z_L as 450 Ω.

16.6　Repeat problem 16.1, with R_S as 25 Ω and Z_L as 50 Ω.

16.7　Refering to Figure 16.10, if R_S is 50 Ω then determine:

(a)　the value of the load resistance if the reflected voltage is 2 V;
(b)　the load resistance if the reflected voltage is –2V.

16.8　A 50 Ω transmission line is used to provide power to two 100 Ω loads, as shown in Figure 16.12. It is decided to change the position of the loads. Two extension cables are inserted each of which are 300 metres long. The characteristic impedance of these cables are 200 Ω and 400 Ω.

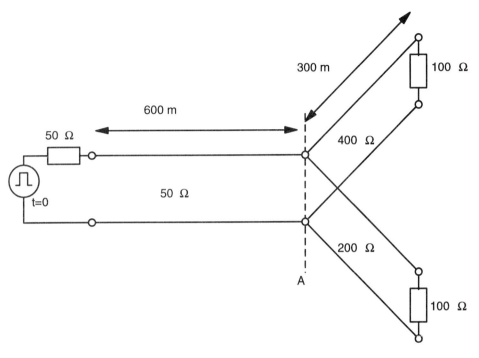

Figure 16.12   Problem Q16.11

Assuming that the lines are lossless and the pulses travel at the speed of light calculate:

(i)   the time taken for the pulse to reach junction A and also to the loads;
(ii)   the pulses on the line at t=2.5 μS;
(iii) the pulses on the line at t=3.5 μS;
(iv)   the value of resistor required at junction A and the resistors required at the loads so that no reflections occur.

16.9   In a digital transmission system two coaxial lines with character-istic impedance 50 Ω have been joined by a line with characteris-tic impedance of 75 Ω, as shown in Figure 16.13. Assuming a 5 V pulse has been sent by the transmitter and that pulses travel at the speed of light, calculate:

(i)   the amplitude of the first two pulses at the receiver and the delay between them;

(ii)   the amplitude of the first two pulses reflected back to the transmitter and the delay between them;

(iii)   the values of resistors required to prevent pulse reflections.

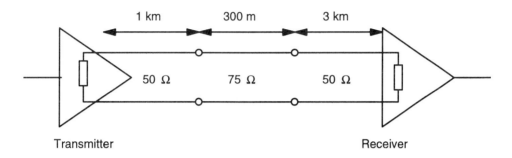

Figure 16.13   Problem 16.9

17

Optical fibre systems

17.1 INTRODUCTION

One the greatest revolutions in data communications is the usage of light waves to transmit digital pulses through fibre optic cables. A light carrying system has an almost unlimited information capacity. Theoretically, it has more than 200 000 times the capacity of a satellite TV system.

Optoelectronics is the branch of electronics which deals with light. Electronic devices that use light operate within the optical part of the electromagnetic frequency spectrum, as shown in Figure 17.1. There are three main bands in the optical frequency spectrum, these are:

- infra-red - the band of light wavelengths that are too long to been seen by the human eye;
- visible - the band of light wavelengths that the human eye responds to;
- ultra-violet - band of light wavelengths that are too short for the human eye to see.

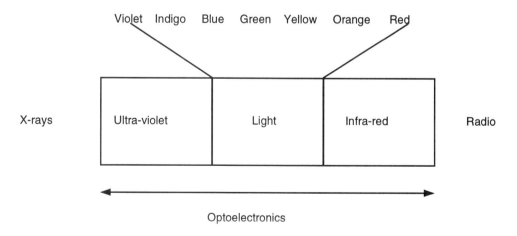

Figure 17.1   EM optoelectronics spectrum

17.2 LIGHT PARAMETERS

17.2.1 Wavelength

Wavelength is defined as the physical distance between two successive points of the same electrical phase. Figure 17.2 shows a wave and its wavelength.

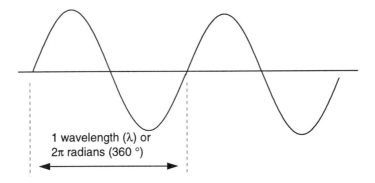

1 wavelength (λ) or
2π radians (360 °)

Figure 17.2   Wavelength of wave

The wavelength is dependent upon the frequency of the wave f, and the velocity of light, c (3×10^8 m.s^{-1}) and is given by:

$$\lambda = \frac{c}{f}$$

The optical spectrum ranges from wavelengths of 0.005 mm to 4000 mm. In frequency terms these are extremely large values from 6×10^{16} Hz to 7.5×10^{10} Hz. It is thus much simpler to talk in terms of wavelengths rather than frequencies.

17.2.2 Colour

The human eye sees violet at one end of the colour spectrum and red on the other. In-between, the eye sees blue, indigo, green, yellow and orange. Two beams of light that have the same wavelength are seen as the same colour and the same colours usually have the same wavelength. Figure 17.3 shows the colour spectrum.

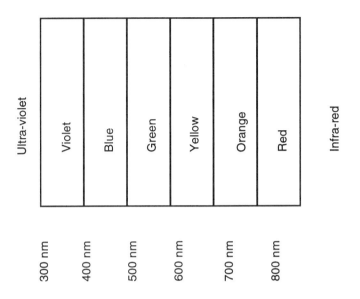

Figure 17.3 Colour spectrum

17.2.3 Velocity of Propagation

In free space electromagnetic waves travel at approximately 300 000 000 m.s^{-1} (186 000 miles.sec^{-1}). However, their velocity is lower when they travel through denser materials. When travelling from a material to an another which is less dense then the light ray to be refracted (or bent) away from the normal. This is shown in Figure 17.4.

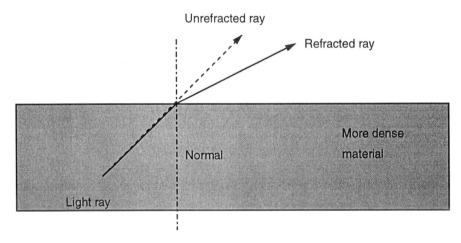

Figure 17.4 Refracted ray

17.2.4 Refractive index

The amount of bending or refraction at the interface between two materials of different densities depends on the refractive index of the two materials. This index is the ratio of the velocity of propagation of a light ray in free space to the velocity of propagation of a light ray the material, as given by:

$$n = \frac{c}{v}$$

where c is speed of light in free space and v is the speed of light in a given medium. Typical refractive indexes are given in Table 17.1.

Table 17.1: Refractive index of sample materials

Medium	Refractive index
Air	1.0003
Water	1.33
Glass Fibre	1.5-1.9
Diamond	2.0-2.42
Gallium Arsenide	3.6
Silicon	3.4

17.3 LIGHT EMITTING DIODE (LED)/ INJECTION LASER DIODE (ILD)

An LED converts electrical energy into light energy. Figure 17.5 shows a simple bias circuit with the voltage applied to the LED by a DC source. It also contains a load resistor to limit the current in the LED. This resistor is determined by subtracting the diode ON voltage from the supply voltage and dividing by the required diode current.

Normally a current of around 10 mA is necessary to produce a good intensity of light. For a GaAs diode the ON voltage is around 2 V. Thus for a GaAs diode, with an ON bias current of 10 mA and a 5 V source, the limiting resistors value would be:

$$R = \frac{5-2}{10 \times 10^{-3}} = 300 \ \Omega$$

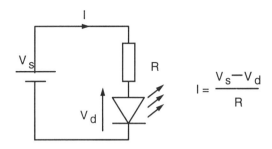

Figure 17.5 Driving an LED

The wavelength of the light emitted depends upon the type of semiconductor used. Gallium Arsenide (GaAs) emits a wavelength in the infra-red range and is typically used as a source of infrared light.

Although the basic GaAs emits infra-red, it can be doped with other materials to provide a wider range of wavelengths. Gallium phosphide (GaP) emits green light and it can radiate a red light depending on the doping.

Gallium arsenide phosphide (GaAsP) emits light over an orange-red range depending on the amount of GaP in the material. With the correct amount of GaP in the material a yellow light is emitted.

Another type of source of light used in optoelectronics is the laser (Light Amplification by Stimulation Emission of Radiation). An injection laser diode (ILD) is an electronic laser which emits light of a single wavelength, known as monochromatic light. ILDs have advantage over LEDs because:

- they produce a more focused light;
- their output radiation power is greater than for an LED, typically 5 mW for ILD and 0.5 mW for an LED;
- they offer higher bit rates as they can be turned ON and OFF faster;
- they produces monochromatic light.

The main disadvantages of ILDs are that they are more expensive, have a shorter lifetime and are more temperature dependent.

17.4 PHOTODIODES

Photodiodes and phototransistors convert light energy (photons) into electrical energy. Their operation is based on the fact that the number of free electrons generated in a semiconductor material is proportional to the intensity of the incident light.

A photodiode must be reverse biased, the reverse biased current varies as the amount of light on the diode junctions. A basic biasing circuit for a photodiode is shown in Figure 17.6. The amount of current is normally extremely small, possibly just a few hundred µA.

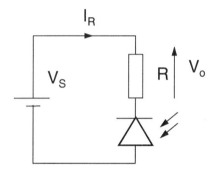

Figure 17.6   Basic biasing arrangement

17.5 FIBRE OPTICS

17.5.1 Introduction

Optical fibres are transparent, dielectric cylinders surrounded by a second transparent dielectric cylinder. Light is transported by a series of reflections from wall to wall from the interface between a core (inner cylinder) and its cladding (outer cylinder). A cross-section of a fibre is given in Figure 17.7.

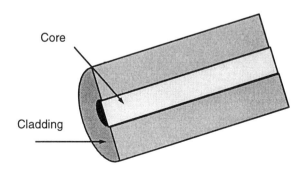

Figure 17.7   Cross-section of an optical fibre

Reflections occur because the core has a higher reflective index than the cladding (it thus has a higher density). Abrupt differences in the refractive index causes the light wave to bounce from the core/cladding interface back through the core to its opposite wall. Thus the light is transported from a light source to a light detector at the other end of the fibre.

17.5.2 Theory

Optical fibres transmit light by total internal reflection (TIR). Light rays passing between the boundaries of two optically transparent media of different densities experience refraction, as shown in Figure 17.8. This changed direction can be determined according to Snells Law:

$$n_1 \sin\theta_1 = n_2 \sin\theta_2$$

Thus

$$\theta_2 = \sin^{-1}\left[\frac{n_1}{n_2}\sin\theta_1\right]$$

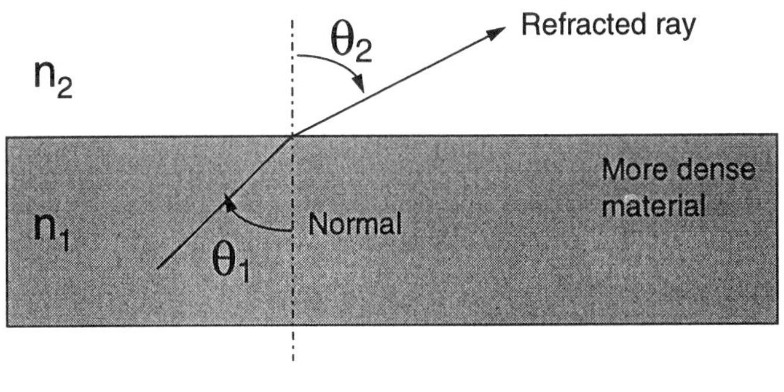

Figure 18.8 Refraction

The angle at which the ray travels along the interface between the two materials is called the critical angle (θ_c) and is given by:

$$\theta_2 = 90°$$

$$\theta_c = \sin^{-1}\left[\frac{n_2}{n_1}\right]$$

It can be shown that when the angle of incident (θ_i) is

$$\theta_i < \sin^{-1}\left[\sqrt{n_1^2 - n_2^2}\right]$$

then the ray is totally reflected from the outer cladding. It then propagates along the fibre being reflected by the cladding on the way, as shown in Figure 17.9. This angle is called the acceptance angle.

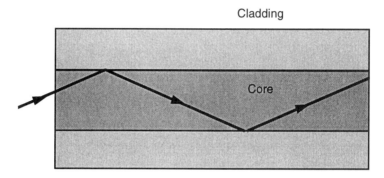

Figure 18.9   Light propagating in an optical fibre

17.5.3 Losses in fibre optic cables

Fibre optic losses result in a lower transmitted light power. This reduces the system bandwidth, information transmission rate, efficiency and overall system capacity. The main losses are:

- Absorption losses - impurities in the glass fibre cause the transmitted wave to be absorbed and converted into heat;
- Material scattering - extremely small irregularities in the structure of the cable causes light to be diffracted. This causes the light to disperse or spread out in many directions. A greater loss occurs a visible wavelengths than at infra-red;
- Chromatic distortion - this is caused by each wavelength of light travelling at differing speeds. They thus arrive at the receiver at different times causing a distorted pulse shape. Monochromatic light reduces this type of distortion;
- Radiation losses - this is caused by small bends and kinks in the fibre which scatters the wave.

- Modal dispersion - is caused by light taking different paths through the fibre. This will each have a different propagation time to travel along the fibre. These different paths are described as modes. Figure 17.10 shows two rays taking different paths. Ray 2 will take a long time to get to the receiver than ray 1.

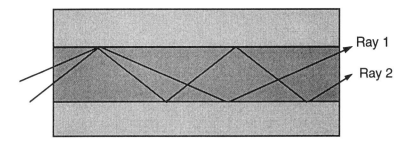

Figure 17.10   Light propagating in different modes

- Coupling losses - these losses are due to light being lost at mismatches at terminations between fibre/fibre, LED/fibre, etc.

17.5.4 Fibre optic link

There are three basic parts to a fibre optic system, the transmitter, the receiver and the fibre guide. The transmitter consists of an analogue-to-digital converter, a voltage-to-current converter, a light source, and a source-to-fibre light coupler. A fibre guide is either ultra-pure glass or plastic cable. The receiver has a fibre-to-light detector coupling device, a photo-detector, a current-to-voltage converter, an amplifier, and an digital-to-analogue converter, as shown in Figure 17.11.

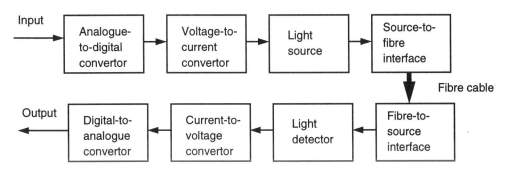

Figure 17.11   Fibre optic communications link

The light source is either an light emitting diode (LED) or a injection laser diode (ILD) and the amount of light emitted by either an LED or ILD is proportional to the current applied.

17.6 TYPICAL OPTICAL FIBRE CHARACTERISTICS

Table 17.2 shows two typical fibre optic cable characteristics. It can be seen that the inside core and the cladding diameters are relatively small i.e. fractions of a millimetre. Normally the cladding is covered in a coating which is then covered in a jacket. These give the cable mechanical strength and also makes it easier to work with. In the case of the 50/125 μm glass cable in Table 17.2 the outer diameter of the cable is 3.2 mm but the inner core diameter is 50 μm.

Normally glass fibre cables have better electrical characteristic over plastic equivalents, but are more prone to breakage and damage. It can be seen that the glass cable has improved bandwidth and lower attenuation over the plastic equivalent.

Table 17.2 Typical fibre optic cables characteristics

	50/125 μm glass	*200 μm PCS*
Construction	Glass	Plastic coated silica (PCS)
Core diameter	50 μm	200 μm
Cladding diameter	125 μm	389 μm
Coating diameter	250 μm	600 μm
Jacket material	Polyethylene	PVC
Overall diameter	3.2 mm	4.8 mm
Connector	9 mm SMA	9 mm SMA
Bandwidth	400 MHz.km^{-1}	25 MHz.km^{-1}
Minimum bend radius	30 mm	50 mm
Temperature range	−15 °C to +60 °C	−10 °C to +50 °C
Attenuation @820 nm	3 dB.km^{-1}	7 dB.km^{-1}

17.7 ADVANTAGES OF FIBRE OPTICS OVER COPPER CONDUCTORS

There are many advantages in using fibre optics cable and very few disadvantages. A summary of the advantages are:

- fibre systems have a greater capacity due to the inherently larger bandwidths available with optical frequencies. Metallic cables contain capacitance and inductance along their conductors, which cause them to act like low-pass filters and limit a signals bandwidths and also its speed of propagation;
- fibre systems are immune from cross-talk between cables caused by magnetic induction. Glass fibres are non-conductors of electricity and therefore do not have a magnetic field associated with them. In metallic cables, the primary cause of cross-talk is magnetic induction between conductors located near each other;
- fibre cables do not suffer from static interference caused by lightning, electric motors, fluorescent lights, and other electrical noise sources. This immunity is because fibres are non-conductors of electricity;
- fibre systems have greater electrical isolation thus allow equipment greater protection from damage due to external sources. For example if the receiver is hit by lightning pulse them it may damage the opto-receiver but a high voltage pulse cannot travel along the optical cable and damage sensitive equipment as the source. They also prevent electrical noise travel back from the receiver to the transmitter, as illustrated in Figure 17.12;

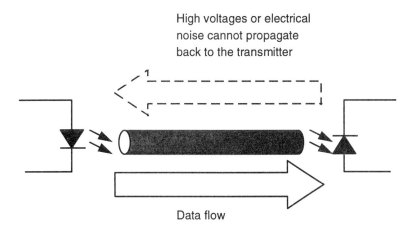

Figure 17.12 Fibre optic isolation

- fibre cables do not radiate energy and therefore cannot cause interference with other communications systems. This characteristic makes fibre systems ideal for military applications, where the effect of nuclear weapons (EMP-electromagnetic pulse interference) has a devastating effect on conventional communications systems;

- fibre cables are more resistive to environmental extremes. They operate over a larger temperature variation than copper cables and are affected less by corrosive liquids and gases;
- fibre cables are safer and easier to install and maintain as glass and plastic as they have no electrical currents or voltages associated with them. Optical fibres can be used around volatile liquids and gases without worrying about the risk of explosions or fires. They are also smaller and more lightweight than copper cables;
- fibre cables are more electrically secure than their copper cables and are virtually impossible to tap into without users knowing about it.

17.8 TUTORIAL

17.1   Determine the wavelength, in nanometers (nm), for the following light frequencies:

   (a) 3.45×10^{14} Hz
   (b) 3.62×10^{14} Hz
   (c) 3.21×10^{14} Hz

17.2   Determine the light frequency for the following wavelengths:

   (a)   670 nm
   (b)   7800 nm
   (c)   710 nm

17.3   For a glass (n=1.5)/quartz (n=1.38) interface and an angle of incidence of 35°, determine the angle of reflection.

17.4   Determine the critical angle and the acceptance angle for the fibre given in question 17.3.

17.5   A fibre optic cable is made up of a glass core (n=1.5) and a fused quartz cladding (n=1.46), determine the critical angle and the minimum acceptance angle.

Appendix A

Practical RS-232 programming

A.1 INTRODUCTION

RS-232 provides an excellent opportunity to investigate practical communications systems. It is a standard device on all new PC and has many practical applications, especially in remote communication, instrumentation and control.

The Intel 8250 IC is commonly used in serial communications. It is mounted onto the PC motherboard or fitted to an I/O card. This section discusses how it is programmed.

A.2 ISOLATED I/O

The RS-232 device(s) map into the PCs isolated I/O memory area, this is separate from the computers main memory. These isolated I/O locations store a single byte and are referred to as ports. Their addresses range from 0000h to FFFFh. Table A.1 shows some of the standard devices which mapped into this memory. For example, the primary serial communications port (COM1:) maps into addresses 3F8h-3FFh.

Table A.1: Typical isolated I/O memory map

Address	Device	Address	Device
000h-01Fh	DMA controller	278h-27Fh	Second parallel port (LPT2:)
020h-03Fh	Interrupt controller	2F8h-2FFh	Second serial port (COM2:)
040h-05Fh	Counter/timer	300h-31Fh	Prototype card
060h-07Fh	Digital I/O	378h-37Fh	Primary parallel port (LPT1:)
080h-09Fh	DMA controller	380h-38Ch	SDLC interface
0A0h-0BFh	NMI reset	3A0h-3AFh	Primary binary synchronous port
0C0h-0DFh	DMA controller	3B0h-3BFh	Monochrome display
0E0h-0FFh	Math co-processor	3D0h-3DFh	Colour/graphics adapter
1F0h-1F8h	AT fixed disk	3F0h-3F7h	Floppy disk controller
200h-20Fh	Game I/O adapter	3F8h-3FFh	Primary serial port (COM1:)
210h-217h	Expansion unit		

A.2.1 Inputting a byte

In Turbo/Borland C/C++, a byte can be read from a port using the `inportb()` function. The general syntax is:

```
value=inportb(PORTADDRESS);
```

> where PORTADDRESS is the address of the input port and `value` is loaded with the 8-bit value from this address. This function is prototyped in the header file `dos.h`.

For Turbo Pascal the equivalent is accessed via the `port[]` array. Its general syntax is as follows:

```
value:=port[PORTADDRESS];
```

> where PORTADDRESS is the address of the input port and `value` the 8-bit value at this address. To gain access to this function the statement `uses dos` needs to be placed near the top of the program.

A.2.2 Outputting a byte

In Turbo/Borland C/C++, a byte can be outputted to a port using the `outportb()` function. The general syntax is

```
outportb(PORTADDRESS,value);
```

> where PORTADDRESS is the address of the output port and `value` is the 8-bit value to be send to this address. This function is prototyped in the header file `dos.h`.

For Turbo Pascal the equivalent is accessed via the `port[]` array. Its general syntax is as follows:

```
port[PORTADDRESS]:=value;
```

where PORTADDRESS is the address of the output port and value is the 8-bit value to be sent to that address. To gain access to this function the statement uses dos requires to be placed near the top of the program.

Note that some C compilers use the functions inp() and outp() instead of inportb() and outportb(). If this is the case either replace all the occurrences of these functions in this appendix with their equivalent or insert the following lines after the header files have been included.

```
#define inportb(portid)       inp(portid)
#define outportb(portid,ch)   outp(portid,ch)
```

This replaces all occurrences of inportb() and outportb() with inp() and outp(), respectively. In C hexadecimal constants are preceded by a 0 (zero) and the character 'x' (0x). In Pascal a dollar sign is used to signify a hex number, for example $C4.

A.3 PROGRAMMING THE SERIAL DEVICE

The main registers used in RS-232 communications are the Line Control Register (LCR), the Line Status Register (LSR) and the Transmit and Receive buffers (see Figure A.1). The Transmit and Receive buffers share the same addresses.

The base address of the primary port (COM1:) is normally set at 3F8h and the secondary port (COM2:) at 2F8h. A standard PC can support up to four COM ports. These addresses are set in the BIOS memory and the address of each of the ports is stored at address locations 0040:0000 (COM1:), 0040:0002 (COM2:), 0040:0004 (COM3:) and 0040:0008 (COM4:). Program A.1 can be used to identify these addresses. The statement:

```
ptr=(int far *)0x0400000;
```

initializes a far pointer to the start of the BIOS communications port addresses. Each address is 16 bits thus the pointer points to an integer value. A far pointer is used as this can access the full 1 MB of memory, a non-far pointer can only access a maximum of 64 kB.

🖹 Program A.1
```
#include <stdio.h>
#include <conio.h>
int  main(void)
{
int  far *ptr; /* 20-bit pointer */

  ptr=(int far *)0x0400000; /* 0040:0000  */ clrscr();

  printf("COM1: %04x\n",*ptr);
  printf("COM2: %04x\n",*(ptr+1));
  printf("COM3: %04x\n",*(ptr+2));
  printf("COM4: %04x\n",*(ptr+3));

  return(0);
}
```

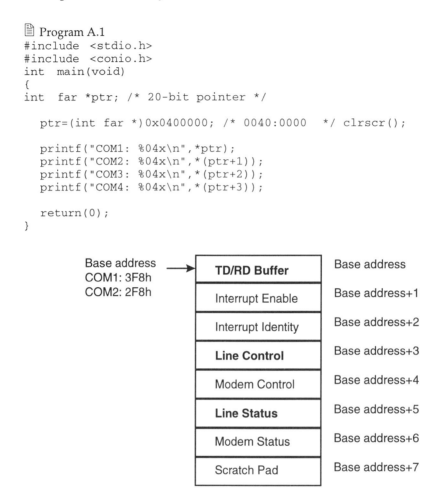

Figure A.1   Serial communication registers

Test run A.1 shows a sample run. In this case there are four COM ports installed on the PC. If any of the addresses is zero then that COM port is not installed on the system.

🖳 Test run A.1
```
COM1:  03f8
COM2:  02f8
COM3:  03e8
COM4:  02e8
```

A.3.1 Line Status Register (LSR)

The LSR determines the status of the transmitter and receiver buffers. It can only be read-from, and all the bits are automatically set by hardware. The bit definitions are given in Figure A.2. When an error occurs in the transmission of a character one (or several) of the error bits is (are) set to a '1'.

One danger when transmitting data is that a new character can be written to the transmitter buffer before the previous character has been sent. This overwrites the contents of the character being transmitted. To avoid this the status bit S_6 is tested to determine if there is still a character still in the buffer. If there is then it is set to a '1', else the transmitter buffer is empty.

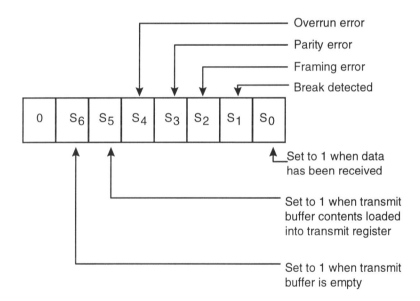

Figure A.2 Line Status Register

The routine to send a character is:

> *Test Bit 6 until set;*
> *Send character;*

The equivalent Turbo Pascal routine is:

```
repeat
    status := port[LSR] and $40;
until (status=$40);
```

When receiving data the S_0 bit is tested to determine if there is a bit in the receiver buffer. To receive a character:

> *Test Bit 0 until set;*
> *Read character;*

The equivalent Turbo Pascal routine is:

```
repeat
    status := port[LSR] and $01;
until (status=$01);
```

Figure A.3 shows how the LSR is tested for the transmission and reception of characters.

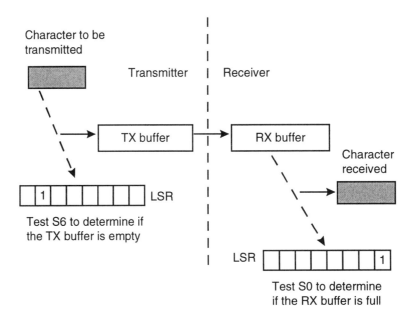

Figure A.3   Testing for the transmission and reception of characters

A.3.2 Line Control Register (LCR)

The LCR sets up the communications parameters. These include the number of bits per character, the parity and the number of stop bits. It can be written to or read from and has a similar function to that of the control registers used in other I/O devices. The bit definitions are given in Figure A.4.

The msb, C_7, must to be set to a '0' in order to access the transmitter and receiver buffers, else if it is set to a '1' the baud rate divider is set up. The baud rate is set by loading an appropriate 16-bit divisor into the addresses of transmitter/receiver buffer address and the next address. The value loaded depends on the crystal frequency connected to the IC. Table A.2 shows divisors for a crystal frequency is 1.8432 MHz. In general the divisor, N, is related to the baud rate by:

$$Baud\ rate = \frac{Clock\ frequency}{16 \times N}$$

For example, for 1.8432 MHz and 9600 baud $N = 1.8432 \times 10^6 / (9600 \times 16)$ = 12 (000Ch).

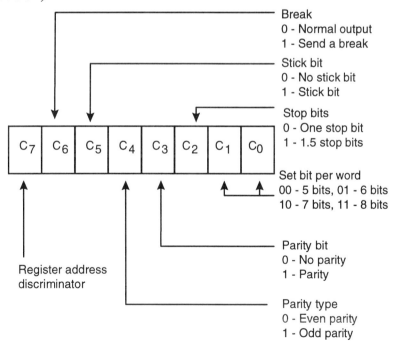

Figure A.4   Line Control Register

Table A.2   Baud rate divisors

Baud rate	Divisor (value loaded into Tx/Rx buffer)
110	0417h
300	0180h
600	00C0h
1200	0060h
1800	0040h
2400	0030h
4800	0018h
9600	000Ch
19200	0006h

A.3.3 Register addresses

The addresses of the main registers are given in Table A.3. To load the baud rate divisor, first the LCR bit 7 is set to a '1', then the LSB is loaded into divisor LSB and the MSB into the divisor MSB register. Finally, bit 7 is set back to a '0'. For example, for 9600 baud, COM1 and 1.8432 MHz clock then 0Ch is loaded in 3F8h and 00h into 3F9h.

When bit 7 is set at a '0' then a read from base address reads from the RD buffer and a write operation writes to the TD buffer. An example is this is shown in Figure A.5.

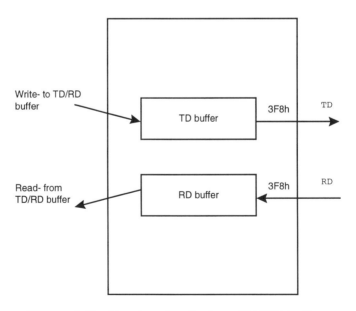

Figure A.5   Read and write-from TD/RD buffer

Table A.3   Serial communications addresses

Primary	Secondary	Register	Bit 7 of LCR
3F8h	2F8h	TD buffer	'0'
3F8h	2F8h	RD buffer	'0'
3F8h	2F8h	Divisor LSB	'1'
3F9h	2F9h	Divisor MSB	'1'
3FBh	2FBh	Line Control Register	
3FDh	2FDh	Line Status Register	

A.3.4 Programming RS232 via DOS

The DOS command mode (or md for DOS Version 6.0) can be used to set the parameters of the serial port. The general format is shown next, options in square brackets ([ ]) are optional.

```
MODE COMn[:]baud[, parity[ ,word_size[ ,stopbits[ ,P]]]]
```

The mode command can also be used for other functions such as setting up the parallel port, text screen, and so on.

```
C:\DOCS\NOTES>mode /?
Configures system devices.
MODE COMm[:][BAUD=b][PARITY=p][DATA=d][STOP=s][RETRY=r]
```

For example

```
C>  mode com2:2400,e,8,1
```

sets up COM2: with 2400 baud, even parity, 8 data bits and 1 stop bit.

```
C>  mode com1:9600
```

changes the baud rate to 9600 on COM1.

A.4 PROGRAMS

Program A.2 uses a loop back on the TD/RD lines so that a character sent by the computer will automatically be received into the receiver buffer and Program A.3 is the Turbo Pascal equivalent.

This set-up is useful in testing the transmit and receive routines. The character to be sent is entered via the keyboard. A *CNTRL-D* (^D) keystroke exits the program. Figure A.6 shows system connections for a 9- and 25-pin connector.

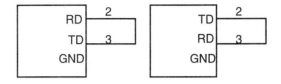

9-pin D-type connector   25-pin D-type connector

Figure A.6   System connections

Program A.2
```c
/* This program transmits a character from COM1: and receives */
/* it via this port. The TD is connected to RD.              */

#define   COM1BASE   0x3F8
#define   COM2BASE   0x2F8
#define   TXDATA     COM1BASE
#define   LCR        (COM1BASE+3)  /*  0x3FB line control    */
#define   LSR        (COM1BASE+5)  /*  0x3FD line status     */

#include <conio.h> /* required for getch()                    */
#include <dos.h>   /* required for inportb() and outportb() */
#include <stdio.h>

/* Some ANSI C prototype definitions  */
void    setup_serial(void);
void    send_character(int ch);
int     get_character(void);

int     main(void)
{
int     inchar,outchar;

   setup_serial();
   do
   {
     puts("Enter char to be transmitted (Cntrl-D to end)");
     outchar=getch();
     send_character(outchar);
     inchar=get_character();
     printf("Character received was %c\n",inchar);
   } while (outchar!=4);
   return(0);
}

void    setup_serial(void)
{
   outportb( LCR, 0x80);
   /* set up bit 7 to a 1  to set Register address bit */

   outportb(TXDATA,0x0C);
   outportb(TXDATA+1,0x00);
```

```
   /* load TxRegister with 12, crystal frequency is 1.8432MHz */

   outportb(LCR, 0x0A);
   /* Bit pattern loaded is 00001010b, from msb to lsb these are:   */
   /* 0 - access TD/RD buffer , 0 - normal output                   */
   /* 0 - no stick bit , 0 - even parity                            */
   /* 1 - parity on, 0 - 1 stop bit                                 */
   /* 10 - 7 data bits                                              */
}

void    send_character(int ch)
{
char    status;
   do
   {
     status = inportb(LSR) & 0x40;
   } while (status!=0x40);
   /*repeat until Tx buffer empty ie bit 6 set*/

   outportb(TXDATA,(char) ch);
}

int  get_character(void)
{
int  status;
   do
   {
     status = inportb(LSR) & 0x01;
   } while (status!=0x01);
   /* Repeat until bit 1 in LSR is set */

   return( (int)inportb(TXDATA));
}
```

📄 ProgramA.3

```
program RS232_1(input,output);
{    This program transmits a character from COM1: and receives }
{    it via this port. The TD is connected to RD.               }
uses crt;
const   TXDATA =    $3F8;   LSR   =    $3FD;
        LCR    =    $3FB;   CNTRLD =   #4;
var     inchar, outchar:char;

procedure   setup_serial;
begin
   port[LCR] := $80;  { set up bit 7 to a 1                     }
   port[TXDATA] := $0C;
   port[TXDATA+1] := $00;
   { load TxRegister with 12, crystal frequency is 1.8432 MHz   }
   port[LCR] := $0A
   { Bit pattern loaded is 00001010b, from msb to lsb these are:}
   { Access TD/RD buffer, normal output, no stick bit           }
   { even parity, parity on, 1 stop bit, 7 data bits            }
end;

procedure   send_character(ch:char);
var        status:byte;
begin
```

```
  repeat
     status := port[LSR] and $40;
  until (status=$40);
     {repeat until bit Tx buffer is empty                        }
  port[TXDATA] := ord(ch); {send ASCII code                      }
end;

function    get_character:char;
var         status,inbyte:byte;
begin

  repeat
     status := port[LSR] and $01;
  until (status=$01);
  inbyte := port[TXDATA];
  get_character:= chr(inbyte);
end;
begin
  setup_serial;

  repeat
     outchar:=readkey;
     send_character(outchar);
     inchar:=get_character;
     writeln('Character received was ',inchar);
  until (outchar=CNTRLD);
end.
```

In the next two programs a transmitter and receiver program are used to transmit data from one PC to another. Program A.4 should run on the transmitter PC and Program A.5 runs on the receiver. The cable connections for 9-pin to 25-pin, 9-pin to 9-pin and 25-pin to 25-pin connectors are given in Figure A.7.

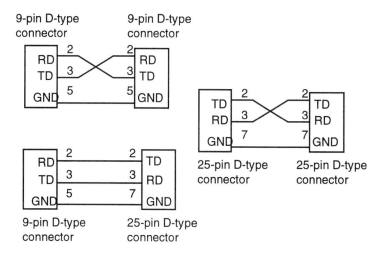

Figure A.7   Connection from one PC to another

📄 Program A.4

```c
/*    send.c                                               */
#define   TXDATA   0x3F8
#define   LSR      0x3FD
#define   LCR      0x3FB

#include <stdio.h>
#include <conio.h>    /* included for getch              */
#include <dos.h>      /* included for inputb and outputb */

void    setup_serial(void);
void    send_character(int ch);

int     main(void)
{
int     ch;

  puts("Transmitter program. Please enter text (Cntl-D to end)");
  setup_serial();
  do
  {
    ch=getche();
    send_character(ch);
  } while (ch!=4);

  return(0);
}

void setup_serial(void)
{

  outportb( LCR, 0x80);
  /* set up bit 7 to a 1  to set Register address bit      */

  outportb(TXDATA,0x0C);
  outportb(TXDATA+1,0x00);
  /* load TxRegister with 12, crystal frequency is 1.8432MHz */

  outportb(LCR, 0x0A);
  /*Bit pattern loaded is 00001010b, from msb to lsb these are:  */
  /*Access TD/RD buffer, normal output, no stick bit            */
  /*even parity, parity on, 1 stop bit, 7 data bits             */
}

void    send_character(int ch)
{
char    status;

  do
  {
    status = inportb(LSR) & 0x40;
  } while (status!=0x40);
  /*repeat until Tx buffer empty ie bit 6 set*/


  outportb(TXDATA,(char) ch);
}
```

📄 Program A.5

```c
/*    receive.c                                                */

#define   TXDATA  0x3F8
#define   LSR     0x3FD
#define   LCR     0x3FB

#include <stdio.h>
#include <conio.h>   /* included for getch               */
#include <dos.h>     /* included for inputb and outputb  */

void    setup_serial(void);
int     get_character(void);


int     main(void)
{
int     inchar;
setup_serial();
   do
   {
     inchar=get_character();
     putchar(inchar);
   } while (inchar!=4);
   return(0);

}
void setup_serial(void)
{
   outportb( LCR, 0x80);
   /* set up bit 7 to a 1  to set Register address bit         */

   outportb(TXDATA,0x0C);
   outportb(TXDATA+1,0x00);
   /* load TxRegister with 12, crystal frequency is 1.8432MHz     */

   outportb(LCR, 0x0A);
   /* Bit pattern loaded is 00001010b, from msb to lsb these are:  */
   /* Access TD/RD buffer, normal output, no stick bit            */
   /* even parity, parity on, 1 stop bit, 7 data bits             */
}

int  get_character(void)
{
int  status;

   do
   {
     status = inportb(LSR) & 0x01;
   } while (status!=0x01);

   /* Repeat until bit 1 in LSR is set */

   return( (int)inportb(TXDATA));
}
```

A.5 USING BIOS

The previous section discussed how the 8250 IC is programmed. Some machines may use a different IC, such as the 8251. An improved method of programming the RS-232 device is to use the BIOS commands. These are device independent and contain programs that can control the RS-232 hardware. The function used, in C, is

```
int bioscom(int cmd,char abyte,int port)
```

where
port   corresponds to the port to use, 0 for COM1:, 1 for COM2:;
cmd   0 – set communications parameters to the value given by abyte;
         1 – send a character;
         2 – receive a character;
         3 – return status of communications.

When cmd is set to 0 the device is programmed. In this mode the definition of the bits in abyte is given in Figure A.8. For example if the function call is bioscom(0,0x42,0) then the RS-232 parameters for COM1: will be 300 baud, no parity, 1 stop bit and 7 data bits.

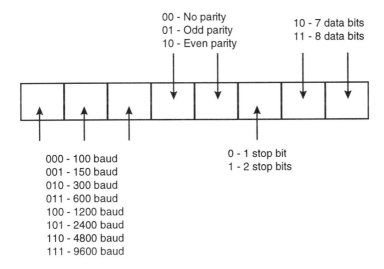

Figure A.8   Bit definitions for bioscom() function

If cmd is set to 3 then the return value from the function is a 16-bit unsigned integer. Bits 8 to 15 are defined as:

> Bit 15 – Time out
> Bit 14 – Transmit shift register empty (character sent)
> Bit 13 – Transmit holding register empty
> Bit 12 – Break detect
> Bit 11 – Framing error
> Bit 10 – Parity error
> Bit 9 – Overrun error
> Bit 8 – Data ready

Program A.6 uses bioscom() to create a transmit/receive program. To test it either loop the TD to the RD or connect two PCs together. The BIOS functions use the RTS and CTS lines in their operation. Thus, connect the RTS to the CTS on the transmitter and receiver or connect the RTS of the transmitter to the CTS of the receiver and the CTS on the transmitter to the RTS of the receiver (see Figure 8.21).

🖹 Program A.6

```
#include <stdio.h>
#include <bios.h>
#include <conio.h>

#define   COM1          0
#define   COM2          1

#define   DATA_READY    0x100
#define   DATABITS7     0x02
#define   DATABITS8     0x03

#define   STOPBIT1      0x00
#define   STOPBIT2      0x04

#define   NOPARITY      0x00
#define   ODDPARITY     0x08
#define   EVENPARITY    0x18

#define   BAUD110       0x00
#define   BAUD150       0x20
#define   BAUD300       0x40
#define   BAUD600       0x60
#define   BAUD1200      0x80
#define   BAUD2400      0xA0
#define   BAUD4800      0xC0
#define   BAUD9600      0xE0
```

```
#define    ESC              0x1B

int     main(void)
{
int     RS232_setting,status,in,ch;

   RS232_setting=BAUD2400 | STOPBIT1 | NOPARITY | DATABITS7;

   bioscom(0,RS232_setting,COM1);

   puts("RS-232 COMBIOS press ESC to exit");

    do
    {

      status = bioscom(3, 0, COM1);

      if (kbhit())
      {
        ch = getch();
        bioscom(1, ch, COM1); /* send character */
      }

      if (status & DATA_READY)
         if ((in = bioscom(2, 0, COM1) & 0x7F) != 0) /* receive char */
            putch(in);

    }  while (ch!=ESC);

    return 0;
}
```

A.6 TUTORIAL

A.1   Write a program that continuously sends the character 'A' to the serial line. Observe the output on an oscilloscope and identify the bit pattern and the baud rate.

A.2   Write a program that continuously sends the characters from 'A' to 'Z' to the serial line. Observe the output on an oscilloscope.

A.3   Complete Table A.4 to give the actual time to send 1000 characters for the given baud rates. Compare these values with estimated values.

   Note that approximately 10 bits are used for each character thus 960 characters/sec will be transmitted at 9600 baud.

A.4 Modify Program A.2 or A.3 so that the program prompts the user for the baud rate when the program is started. A sample run is shown in Test run A.2.

Table A.4 Baud rate timing

Baud rate	Time to send 1000 characters (sec)
110	
300	
600	
1200	
2400	
4800	
9600	
19200	

🖥 Test run A.2

```
Enter baud rate required:
1   110
2   150
3   300
4   600
5   1200
6   2400
7   4800
8   9600
>> 8
RS232 transmission set to 9600 baud
```

A.5 Modify the setup_serial() routine so that the RS-232 parameters can be passed to it. These parameters should include the comport (either COM1: or COM2:), the baud rate, the number of data bits and the type of parity. An outline of the modified function is given in Program A.7.

📄 Program A.7
```
#define   COM1BASE   0x3F8
#define   COM2BASE   0x2F8

#define   COM1       0
#define   COM2       1

enum    baud_rates  {BAUD110,BAUD300,BAUD600,BAUD1200,
                     BAUD2400,BAUD4800,BAUD9600};

enum    parity      {NO_PARITY,EVEN_PARITY,ODD_PARITY};

enum    databits    {DATABITS7,DATABITS8};

#include <conio.h>
```

```c
#include <dos.h>
#include <stdio.h>


/* Some ANSI C prototype definitions  */
void setup_serial(int comport, int baudrate, int parity,
                  int databits);
void send_character(int ch);
int  get_character(void);

int  main(void)
{
int  inchar,outchar;

   setup_serial(COM1,BAUD2400,EVEN_PARITY,DATABITS7);
   ::::::::::::etc.

}


void setup_serial(int comport, int baudrate,
                  int parity, int databits)
{
int  tdreg,lcr;

   if (comport==COM1)
   {
      tdreg=COM1BASE;
      lcr=COM1BASE+3;
   }
   else
   {
      tdreg=COM2BASE;
      lcr=COM2BASE+3;
   }

   outportb( lcr, 0x80);
   /* set up bit 7 to a 1  to set Register address bit  */

   switch(baudrate)
   {

   case BAUD110: outportb(tdreg,0x17);outportb(tdreg+1,0x04);
                 break;
   case BAUD300: outportb(tdreg,0x80);outportb(tdreg+1,0x01);
                 break;
   case BAUD600: outportb(tdreg,0x00);outportb(tdreg+1,0xC0);
                 break;
   case BAUD1200:
      outportb(tdreg,0x00);outportb(tdreg+1,0x40); break;
   case BAUD2400:
      outportb(tdreg,0x00);outportb(tdreg+1,0x30); break;
   case BAUD4800:
      outportb(tdreg,0x00);outportb(tdreg+1,0x18); break;
   case BAUD9600:
      outportb(tdreg,0x00);outportb(tdreg+1,0x0C); break;
   }
      ::::::::::: etc.
}
```

A.6   One problem with Programs 7.4 and 7.5 is that when the return key is pressed only one character is sent. The received character will be a carriage return which returns the cursor back to the start of a line and not to the next line. Modify the receiver program so that a line feed will be generated automatically when a carriage return is received. Note a carriage return is an ASCII 13 and line feed is a 10.

A.7   Modify the `get_character()` routine so that it returns an error flag if it detects an error or if there is a time-out. Table A.5 lists the error flags and the returned error value. An outline of the C code is given in Program 7.10. If a character is not received within 10 seconds an error message should be displayed.

Program A.8

```
#include <stdio.h>
#include <dos.h>

#define   TXDATA   0x3F8
#define   LSR      0x3FD
#define   LCR      0x3FB

void    show_error(int ch);
int     get_character(void);

enum      RS232_errors   {PARITY_ERROR=-1, OVERRUN_ERROR=-2,
          FRAMING_ERROR=-3,  BREAK_DETECTED=-4, TIME_OUT=-5};

int     main(void)
{
int     inchar;
  do
  {
    inchar=get_character();

    if (inchar<0) show_error(inchar);
    else printf("%c",inchar);
  } while (inchar!=4);

  return(0);
}

void    show_error(int ch)
{
  switch(ch)
  {
  case PARITY_ERROR: printf("Error: Parity error/n"); break;
  case OVERRUN_ERROR: printf("Error: Overrun error/n"); break;
```

```
case FRAMING_ERROR: printf("Error: Framing error/n"); break;
case BREAK_DETECTED: printf("Error: Break detected/n");break;
case TIME_OUT: printf("Error: Time out/n"); break;
}
}

int  get_character(void)
{
int  instatus;
  do
  {
     instatus = inportb(LSR) & 0x01;
     if (instatus & 0x02) return(BREAK_DETECTED);
                   :::: etc
  } while (instatus!=0x01 );

  return( (int) inportb(TXDATA) );

}
```

Table A.5   Error returns from get_character()

Error condition	Error flag return	Notes
Parity error	−1	
Overrun error	−2	
Framing error	−3	
Break detected	−4	
Time-out	−5	get_character() should time-out if no characters are received with 10 seconds.

Test the routine by connecting two PCs together and set the transmitter with differing RS232 parameters.

A.7 PROJECTS

A.7.1 Project 1: Half-duplex link

Design and implement a half-duplex link between two computers, that is, only one computer can talk at a time. The same program should run on both computers but one should automatically go into talk mode when a key is pressed on the keyboard and the other as a listener. When the talker transmits an ASCII code 04 (^D) the mode of the computers should swap, that is, the talker should listen and the listener should talk. Figure A.9 shows a sample conversation.

A.7.2 Project 2: Full-duplex link

Design and implement a *simulated* full-duplex link between two computers, that is, both computers can talk and listen at the same time. The same program should run on both computers. *Hint*: one possible implementation is to loop within the `get_character()` routine and break out of it if a character is received from the line or if a character has been entered from the keyboard. If a character is entered from the keyboard it should be sent and the program then returns to the `get_character()` routine.

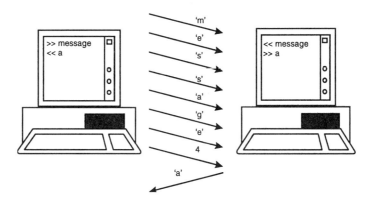

Figure A.9   System operation

A.7.3 Project 3: Simulated software handshaking

Set up two PCs so that one PC transmits characters to the other and the receiving PC displays them. If the space bar is pressed on the receiver PC then it should send an X-OFF character to the transmitter. The transmitter should then display a message informing the user that the receiver is busy. When the spacebar is pressed again on the receiver it should transmit an X-ON character and the transmitter is free to transmit more characters to the receiver. So it continues, with the receiver using the space bar to toggle its busy/idle state.

A.7.4 Project 4: File transfer

Design and implement a program that transfers a file between two PCs. The name of the file should be initially sent following a NULL character (to delimit the filename). Next, the contents of the file are sent and finally an EOF character.

Appendix B

Data communications standards

B.1 STANDARDS

Table B.1 lists some of standards relating to data communications. The CCITT (now known as the ITU) and the ISO are the main international standards organizations. The CCITT standard that relate to the transmission of data over telephone circuits are defined in the V. series, Packet-switched networks standards are defined in the X. series and Integrated Digital Services Network (ISDN) standards are defined in the I. series.

Table B.1: Data communications standards

ISO/CCITT standard	Other standard	Description
	ANSI X3T9.5	FDDI standard
CCITT I.430 CCITT I.431		Physical layer interface to an ISDN network
CCITT I.440 CCITT I.441		Data layer interface to an ISDN network
CCITT I.450/ CCITT I.451		Network layer interface to an ISDN network
CCITT V.10	EIA RS-423	Serial transmission up to 300 kbps/ 1200 m
CCITT V.11	EIA RS-422	Serial transmission up to 10 Mbps/ 1200 m
CCITT V.21		Full-duplex modem transmission at 300 bps
CCITT V.22		Half-duplex modem transmission at 600/1200 bps
CCITT V.22bis		Full-duplex modem transmission at 1200/2400 bps

CCITT V.23		Full-duplex modem transmission at 1200 bps and receive at 75 bps
CCITT V.24	EIA RS-232C	Serial transmission up to 20 kbps/ 20 m
CCITT V.25bis		Modem command language
CCITT V.27		Full-duplex modem transmission at 2400/ 4800 for leased lines
CCITT V.29		Full-duplex modem transmission at 9600 bps over leased lines
CCITT V.32		Full-duplex modem transmission at 4800/9600 bps
CCITT V.32bis		Full-duplex modem transmission at 7200, 12000 and 14400 bps
CCITT V.35	EIA RS-449	CCITT standard for the RS-449 interface
CCITT V.42		Error control protocol
CCITT X.21		Physical layer interface to connection for synchronous transfer on PSDN
CCITT X.25		Connection to a packet-switched network
ISO 8802.4	IEEE 802.2	Token passing in a token ring LAN network
ISO 8802.5	IEEE 802.3	Token ring topology

B.2 INTERNATIONAL ALPHABET NO. 5

ANSI defined a standard alphabet known as ASCII. This has since been adopted by the CCITT as a standard, known as IA5 (International Alphabet No. 5). The following tables define this alphabet in binary, as a decimal, as a hexadecimal value and as a character.

Binary	Decimal	Hex.	Character	Binary	Decimal	Hex.	Character
00000000	0	00	NUL	00010000	16	10	DLE
00000001	1	01	SOH	00010001	17	11	DC1
00000010	2	02	STX	00010010	18	12	DC2
00000011	3	03	ETX	00010011	19	13	DC3
00000100	4	04	EOT	00010100	20	14	DC4
00000101	5	05	ENQ	00010101	21	15	NAK
00000110	6	06	ACK	00010110	22	16	SYN
00000111	7	07	BEL	00010111	23	17	ETB
00001000	8	08	BS	00011000	24	18	CAN
00001001	9	09	HT	00011001	25	19	EM
00001010	10	0A	LF	00011010	26	1A	SUB
00001011	11	0B	VT	00011011	27	1B	ESC
00001100	12	0C	FF	00011100	28	1C	FS
00001101	13	0D	CR	00011101	29	1D	GS
00001110	14	0E	SO	00011110	30	1E	RS
00001111	15	0F	SI	00011111	31	1F	US

Binary	Decimal	Hex.	Character	Binary	Decimal	Hex.	Character
00100000	32	20	SPACE	00110000	48	30	0
00100001	33	21	!	00110001	49	31	1
00100010	34	22	"	00110010	50	32	2
00100011	35	23	£/#	00110011	51	33	3
00100100	36	24	$	00110100	52	34	4
00100101	37	25	%	00110101	53	35	5
00100110	38	26	&	00110110	54	36	6
00100111	39	27	/	00110111	55	37	7
00101000	40	28	(	00111000	56	38	8
00101001	41	29	)	00111001	57	39	9
00101010	42	2A	*	00111010	58	3A	:
00101011	43	2B	+	00111011	59	3B	;
00101100	44	2C	,	00111100	60	3C	<
00101101	45	2D	–	00111101	61	3D	=
00101110	46	2E	.	00111110	62	3E	>
00101111	47	2F	/	00111111	63	3F	?

Binary	Decimal	Hex.	Character	Binary	Decimal	Hex.	Character
01000000	64	40	@	01010000	80	50	P
01000001	65	41	A	01010001	81	51	Q
01000010	66	42	B	01010010	82	52	R
01000011	67	43	C	01010011	83	53	S
01000100	68	44	D	01010100	84	54	T
01000101	69	45	E	01010101	85	55	U
01000110	70	46	F	01010110	86	56	V
01000111	71	47	G	01010111	87	57	W
01001000	72	48	H	01011000	88	58	X
01001001	73	49	I	01011001	89	59	Y
01001010	74	4A	J	01011010	90	5A	Z
01001011	75	4B	K	01011011	91	5B	[
01001100	76	4C	L	01011100	92	5C	\
01001101	77	4D	M	01011101	93	5D	]
01001110	78	4E	N	01011110	94	5E	`
01001111	79	4F	O	01011111	95	5F	_

Binary	Decimal	Hex.	Character	Binary	Decimal	Hex.	Character
01100000	96	60		01110000	112	70	p
01100001	97	61	a	01110001	113	71	q
01100010	98	62	b	01110010	114	72	r
01100011	99	63	c	01110011	115	73	s
01100100	100	64	d	01110100	116	74	t
01100101	101	65	e	01110101	117	75	u
01100110	102	66	f	01110110	118	76	v
01100111	103	67	g	01110111	119	77	w
01101000	104	68	h	01111000	120	78	x
01101001	105	69	i	01111001	121	79	y
01101010	106	6A	j	01111010	122	7A	z
01101011	107	6B	k	01111011	123	7B	{
01101100	108	6C	l	01111100	124	7C	:
01101101	109	6D	m	01111101	125	7D	}
01101110	110	6E	n	01111110	126	7E	~
01101111	111	6F	o	01111111	127	7F	DEL

Appendix C

Data communications connections

C.1 RS-232C INTERFACE

Table C.1: RS-232C connections

9-pin D-type	25-pin D-type	Name	RS-232 name	Description	Signal Direction on DCE
	1		AA	Protective GND	
3	2	TXD	BA	Transmit Data	IN
2	3	RXD	BB	Receive Data	OUT
7	4	RTS	CA	Request to Send	IN
8	5	CTS	CB	Clear to Send	OUT
6	6	DSR	CC	Data Set Ready	OUT
5	7	GND	AB	Signal GND	
1	8	CD	CF	Received line signal detect	OUT
	9		–	RESERVED	–
	10		–	RESERVED	–
	11			UNASSIGNED	–
	12		SCF	Secondary Received Line Signal Detector	OUT
	13		SCB	Secondary Clear To Send	OUT
	14		SBA	Secondary Transmitted Data	IN
	15		DB	Transmission Signal Element Detector	OUT
	16		SBB	Secondary Received Data	OUT
	17		DD	Receiver Signal Element Time	OUT
	18			UNASSIGNED	–
	19		SCA	Secondary Request To Send	IN
4	20	DTR	CD	Data Terminal Ready	IN
	21		CG	Signal Quality Detector	OUT
9	22	RI	CE	Ring Indicator	OUT
	23		CH/CI	Data Signal Rate Selector	IN/OUT
	24		DA	Transmit Signal Element Timing	IN
	25			UNASSIGNED	–

C.2 RS-449 INTERFACE

RS-449 defines a standard for the function/mechanical interface for DTEs/DCEs for serial communications and is usually used with synchronous transmissions. Table C.2 lists the main connections.

Table C.2: RS-449 connections

Pin number	Mnemonic	Description
1		Shield
2	SI	Signalling Rate Indicator
3,21		Spare
4,22	SD	Sending Time
5,23	ST	Receive Data
6,24	RD	Receive Data
7,25	RS	Request to Send
8,26	RT	Receive Timing
9,27	CS	Clear To Send
10	LL	Local Loopback
11,29	DM	Data Mode
12,30	TR	Terminal Ready
13,31	RR	Receiver Ready
14	RL	Remote Loopback
15	IC	Incoming Call
16	SF/SR	Select Frequency/ Signalling Rate Select
17,37	TT	Terminal Timing
18	TM	Test Mode
19	SG	Signal Ground
20	RC	Receive Common
28	IS	Terminal in Service
32	SS	Select Standby
33	SQ	Signal Quality
34	NS	New Signal
36	SB	Standby Indicator
37	SC	Send Common

Appendix D

Ethernet voltages and Fast Ethernet

D.1 ETHERNET VOLTAGE LEVELS

Baseband Ethernet transmits onto a single ether at a rate of 10 Mbps. This gives a bit period of around 100 ns. When the ether is not busy the voltage on the ether is nominally +0.7 V which provides a carrier sense signal for all nodes on the network and is also known as the heartbeat. A low voltage is nominally −0.7 V.

When transmitting, a transceiver unit transmits a preamble of consecutive 1s and 0s. The coding used is Manchester coding which represents a 0 as a high to a low voltage transition and a 1 as a low to high transition. Thus when the preamble is transmitted the voltage changes between +0.7 and −0.7 V, as illustrated in Figure D.1. If after the transmission of the preamble no collisions are detected then the rest of the frame is sent.

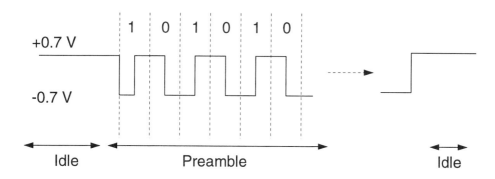

Figure D.1   Ethernet digital signal

Unfortunately, the ether and transceiver electronics are not perfect. The transmission line contains resistance and capacitance which distorts the shape of the pulse. There is also a delay period in the time that a pulse takes to travel along the line to all the nodes on a segment. The number of transceivers which connect to the segment also has an effect on the electrical loading on the ether. Figure D.2 shows a practical measurement of the voltages on an Ethernet segment with coaxial cable (that is, 10Base2). It shows that the pulses are rounded and do not have square edges, as the pulses in Figure D.1. Measurements also show that the idle voltage on the line is approximately +1 V and the pulse amplitude is ±0.8 V.

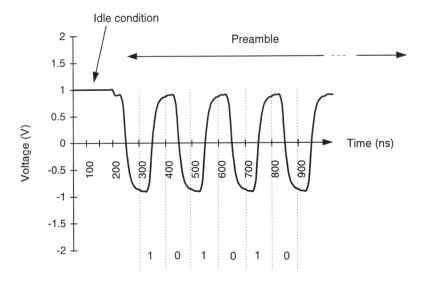

Figure D.2   Practical Ethernet voltages

The start of frame delimiter is identified by two consecutive logic 1s, this is shown in Figure D.3.

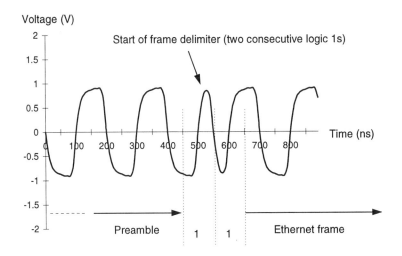

Figure D.3   Start of frame delimiter

D.2 FAST ETHERNET AND 100VG-ANYLAN

Standard 10 Mbps Ethernet does not performance well when many users running multi-media applications. Two improvements to the standard are Fast Ethernet and 100VG-AnyLAN. The IEEE have defined standards for these, IEEE 802.3u for Fast Ethernet and 802.12 for 100VG-AnyLAN. Both are supported by many manufacturers and use a bit rates of 100 Mbps. This gives, at least, ten times the performance of standard Ethernet.

D.2.1 Fast Ethernet

Fast Ethernet, or 100BASE-T, is simply 10BASE-T running at ten times the bit rate. It is a natural progression from standard Ethernet and thus allows existing Ethernet networks to be easily upgraded. Unfortunately, as with standard Ethernet, nodes contend for the network which reduces the network efficient when there are high traffic rates. Also, as it uses collision detect, the maximum segment length is limited by the amount of time for the farthest nodes on a network to properly detect collisions. On a Fast Ethernet network with twisted-pair copper cables this distance is 100 m and for a fibre optic link this is 400 m.

D.2.2 100VG-Any LAN

The 100VG-AnyLAN standard (IEEE 802.12) was developed mainly by Hewlett Packard and overcomes the contention problem by using a priority based, round robin arbitration method, known as Demand Priority Access Method (DPAM). Unlike Fast Ethernet, nodes always connect to a hub which regularly scans its input ports to determine it any nodes have requests pending.

It has an in-built priority mechanism with two priority levels: a high priority request and a normal priority request. A normal priority request is used for non-real time data, such as data files, and so on. High priority requests are used for real-time data, such as speech or video data. At present there is limited usage of this feature hub and there is no support mechanism for this facility after the data has left the hub.

100VG-AnyLAN allows up to seven level hubs (that is, one root and six cascaded hubs) with a maximum distance of 150 m between nodes.

Unlike, other forms of Ethernet it allows any number of node to be connected to a segment.

D.2.3 Migration from Ethernet to Fast Ethernet or 100VG-AnyLAN

If an existing network is based on standard Ethernet then, in most cases, the best network upgrade is either to Fast Ethernet or 100VG-AnyLAN. Since the protocols and access methods are the same there is no need to change any of network management software or application programs. The upgrade path for Fast Ethernet is simple and could be:

- Upgrade high data rate nodes, such as servers or high powered workstations to Fast Ethernet;
- Gradually upgrade NICs (Network Interface Cards) on Ethernet segments to cards which support both 10BASE-T and 100BASE-T. These cards automatically detect the transmission rate to give either 10 or 100 Mbps.

The upgrade path to 100VG-AnyLAN is less easy as it relies on hubs and, unlike Fast Ethernet, most NICs have different network connectors, one for 10BASE-T and the other for 100VG-AnyLAN (although it is likely that more NICs will have automatic detection). A possible path could be:

- Upgrade high data rate nodes, such as servers or high powered workstations to 100VG-AnyLAN;
- Install 100VG-AnyLAN hubs;
- Connect nodes to 100VG-AnyLAN hubs and change-over connectors.

D.2.4 Network performance

It is difficult to assess the performance differences between Fast Ethernet and 100VG-AnyLAN. Fast Ethernet uses a well proven technology but suffers from network contention. 100VG-AnyLAN is a relatively new technology and the handshaking with the hub increases delay time.

Glossary

Address

A unique label for the location of data or the identity of a communications device.

Address Resolution Protocol (ARP)

A TCP/IP process which maps an IP address to an Ethernet address.

American National Standards Institute (ANSI)

ANSI is a non-profit organization which is made up of expert committees that publish standards for national industries.

American Standard Code for Information Interchange (ASCII)

An ANSI-defined character alphabet which has since been adopted as a standard international alphabet for the interchange of characters.

Amplitude modulation (AM)

Information is contained in the amplitude of a carrier.

Amplitude-Shift Keying (ASK)

Uses two, or more, amplitudes to represent binary digits. Typically used to transmit binary data over speech-limited channels.

Application layer

The highest layer of the OSI model.

Asynchronous transmission

Transmission where individual characters are sent one-by-one. Normally each character is delimited by a start and stop bit. With asynchronous communication the transmitter and receiver only have to be roughly synchronized.

Bandwidth

The range of frequencies contained in a signal. As an approximation it is the difference between the highest and lowest frequency in the signal.

Baseband

Data transmission using unmodulated signals.

Baud rate

The number signalling elements sent per second with a RS-232, or modem, communications. In RS-232 the baud rate is equal to the bit-rate. With modems, two or more bits can be encoded as a single signalling element, such as 2 bits being represented by four different phase shifts (or one signalling element). The signalling element could change its amplitude, frequency or phase-shift to increase the bit-rate. Thus the bit-rate is a better measure of information transfer.

Bit stuffing

The insertion of extra bits to stop the appearance of a defined sequence. In HDLC the bit sequence 01111110 delimits the start and end of a frame. Bit stuffing stops this bit sequence from occurring anywhere in the frame by the receiver inserting a 0 whenever their are five consecutive 1's transmitted. At the receive if five consecutive 1's are followed by a 0 then the 0 is deleted.

Bridge

A device which physically links two or more networks using the same communications protocols, such as Ethernet/ Ethernet or token ring/ token ting.

Broadband

Data transmission using multiplexed data using an analogue signal or high-frequency electromagnetic waves.

Buffer

A temporary-storage space in memory.

Bus

A network topology where all nodes share a common transmission medium.

Byte

A group of eight bits, see octet.

Carrier Sense Multiple Access/ Carrier Detect (CSMA/CD)

A network where all nodes share a common bus. Nodes must contend for the bus and if a collision occurs then all colliding node back-off for a random time period.

CCITT

The Consultative Committee for International Telephone and Telegraph (now known at the ITU-TSS) is an advisory committee established by the United Nations. They attempt to establish standards for inter-country data transmission on a wide-wide basis.

Checksum

An error-detection scheme in which bits are grouped to form integer values and then each of the value is summated. Normally, the negative of this value is then added as a checksum. At the receiver, all the grouped values and the checksum are summated and, in the absence of errors, the result should be zero.

CRC

Cyclic Redundancy Check. An error-detection scheme. Used in most HDLC-related data link applications.

Cross-talk

Interference noise caused by conductors radiating electromagnetic radiation to couple into other conductors.

Data Communications Equipment (DCE)
Devices which establish, maintain and terminate a data communications conversation.

Data Terminal Equipment (DTE)
Devices at the end of the data communications connection.

Digital modulation
Method of converting digital data into a form which can be transmitted over a band-limited channel. Methods use either ASK, FSK, PSK or a mixture of ASK, FSK and PSK.

Direct Distance Dialling (DDD)
Allows modems to communicate directly without going through operator services.

Distributed system
A computer system in which computing, storage and other resources are dispersed throughout a network.

Electronic Industries Association (EIA)
US standards organization specializing in electrical interfaces.

Ethernet
A local area network which uses coaxial, twisted-pair or fibre optic cable as a communication medium. It transmits at a rate of 10 Mbps and was developed by DEC, Intel and Xerox Corporation. The IEEE 802.3 network standard is based upon Ethernet.

Ethernet address
A 48-bit number that identifies a node on an Ethernet network. Ethernet addresses are assigned by the Xerox Corporation.

Even parity
An error-detection scheme where defined bit-grouping have an even number of 1's.

Extended Binary Coded Decimal Interchange Code (EBCDIC)
An 8-bit code alphabet developed by IBM allowing 256 different bit patterns for character definitions.

Fibre Distributed Data Interface (FDDI)
The specification for a high-speed fibre-optic ring network.

File Transfer Protocol (FTP)
A protocol for transmitting files between host computers using the TCP/IP protocol.

Flow control	Procedure to regulate the flow of data between two nodes.
Frequency Modulation (FM)	Information is contained in the frequency of a carrier.
Frequency-Division Multiplexing (FDM)	Simultaneous transmission of several information channels using different frequencies for each channel.
Frequency-shift Keying (FSK)	Uses two, or more, frequencies to represent binary digits. Typically used to transmit binary data over speech-limited channels.
Full-Duplex (FDX)	Simultaneous, two-way communications.
Gateway	A device that connects networks using different communications protocols, such as Ethernet/FDDI or Ethernet/ token ring. It provides protocol translation, in contrast to a bridge which connects two networks that are of the same protocol.
Half-duplex (HDX)	Two-way communications, one at a time.
Handshaking	A reliable method for two devices to pass data.
High-Level Data Link Control (HDLC)	ISO standard for the data link layer.
Host	A computer that communicates over a network. A host can both initiate communications and respond to communications that are addressed to it.
Hub	A hub is a concentration point for data and repeats data from one node to all other connected nodes.
IEEE 802.2	A set of IEEE-defined specifications for Logical Link Control (LLC) layer. It provides some network functions and interfaces the IEEE 802.5, or IEEE 802.3, standards to the transport layer.
IEEE 802.3	A set of IEEE-specifications for CSMA/CD networks. It was developed by the IEEE 802.3 committee and has since been adopted by ANSI. Its specifications includes network protocol and hardware specifications.

IEEE 802.4   Token bus specifications.

IEEE 802.5   Token ring specifications.

Institute of Electrical and Electronic Engineers (IEEE)   An international professional society which issues standards.

International Telegraph Union Telecommunications Standards Sector (ITU-TSS)   Organization which has replaced the CCITT.

Internet   Connection of nodes on a global network which use a DARPA-defined Internet address.

internet   Two or more connected networks that may, or may not, use the same communication protocol. The device that connects the networks may be a router, bridge or a gateway.

Internet address   An address that conforms to the DARPA-defined Internet protocol. A unique, four byte number identifies a host or gateway on the Internet. This consists of a network number followed by a host number. The host number can be further divide into a subnet number.

Internet addresses are normally expressed as four decimal numbers, ranging between 0-255, separated by periods.

ISO   International Standards Organization

Leased line   A permanent telephone line connection reserved exclusively by the leased customer. There is no need for any connection and disconnection procedures.

Light-emitting diode (LED)   A device which converts electrical current into light.

Line driver   A device which converts an electrical signal to a form that is transmittable over a transmission line. Typically, it provides the required power, current and timing characteristics.

Link layer   Layer 2 of the OSI model.

Link segment   A point-to-point link terminated on either side by a repeater. Nodes can not be attached to a link segment.

Local Area Network (LAN)   A data communications system that allows a number of independent devices to communicate.

Logical Link Control (LLC)   see IEEE 802.2

Manchester coding   Digital coding technique where each bit period is divided in a positive and a negative half. This helps to embed timing information in the transmitted signal.

Media Access Control (MAC)   Media-specific access-control for token ring and Ethernet.

Media Interface Controller (MIC)   The connector used to attach dual-attachment FDDI stations to a fibre-optic ring.

Medium Attachment Unit (MAU)   A device which connects nodes to the IEEE 802.3 network.

Modem (Modulator-Demodulator)   A device which converts binary digits into a form which can be transmitted over a speech-limited transmission channel.

Network Architecture   The organization of communications devices and their interconnection.

Network controller   In Ethernet, it is device which passes bit frames from the network and the local memory of the computer. Coupled with a network transceiver, it also handles signal processing, encoding and network media access.

Node   Any point in a network which provides communications services or where devices interconnect.

Octet   Same as a byte, a group of eight bits (typically used in communications terminology)

Odd parity   An error-detection scheme where a defined bit-grouping has an even number of 0's.

Optical Repeater	A device that receives, restores, and re-times signals from one optical fibre segment to another.
Packet	A sequence of binary digits that is transmitted as a unit in a computer network. A packet usually contains control information and data.
Phase-Locked Loop (PLL)	Converts FM signals, with a certain range, back into the unmodulated signal. The range at which the PLL 'picks-up' the FM signal is known as the capture range.
Phase-Shift Keying (PSK)	Uses two, or more, phase-shifts to represent binary digits. Typically used to transmit binary data over speech-limited channels.
Protocol	A specification for coding of messages exchanged between two communications processes.
Repeater	A device that receives, restores, and re-times signals from one segment of a network and passes them on to another. Both segments must have the same type of transmission medium and share the same set of protocols. A repeater cannot translate protocols.
RJ-45	Connector used with US telephones and with twisted-pair cables. It is also used in ISDN networks.
Routing node	A node that transmits packets between similar networks. A node that transmits packets between dissimilar networks is called a gateway.
RS-232C	EIA-defined standard for serial communications.
RS-422, 423	EIA-defined standard which uses a higher transmission rates and cable lengths than RS-232.
RS-449	EIA-defined standard for the interface between a DTE and DCE for 9- and 37-way D-type connectors.
RS-485	EIA-defined standard which is similar to RS-422 but uses a balanced connection.
Segment	A segment is any length of LAN cable terminated at both

ends. In a bus network, segments are electrically continuous pieces of the bus, connected at by repeaters.

Simplex   One-way communication.

Synchronous   Transmission where the transmitter and receiver
transmission   synchronize their to each other. The bits are sent at a fixed rate which means that no start and stop bits are required (see asynchronous communications)

TCP/IP Internet   An Internet is made up of networks of nodes that can communicate with each other using TCP/IP protocols.

Time-Division   Simultaneous transmission of several information channels
Multiplexing (TDM)   using different time-slots for each channel.

Token   A token transmits data around a token ring network.

Topology   The physical and logical geometry governing placement of nodes on a network.

Transceiver   A device that transmits and receives signals.

Transmission Control   A standard protocol, defined by the Defence Advanced
Protocol and Internet   Research Projects Agency (DARPA), to allow different
Protocol (TCP/IP)   host computers to communicate over a variety of network types.

V.24   CCITT-defined specification, similar to RS-232C.

World-Wide Web   The interconnection of networks on the Internet.
(WWW)

X-ON/ X-OFF   The Transmitter On/ Transmitter Off characters are used to control the flow of information between two nodes.

X.21   CCITT-defined specification for the interconnection of DTEs and DCEs for synchronous communications.

X.25   CCITT-defined specification for packet-switched network connections.

Common abbreviations

AA	auto answer
ABM	asynchronous balanced mode
AC	access control
ACK	acknowledge
ADC	analogue-to-digital converter
ADPCM	adaptive pulse code modulation
AFI	authority and format identifier
AM	amplitude modulation
AMI	alternative mark inversion
ANSI	American National Standard Institute
ARM	asynchronous response mode
ASCII	American standard code for information exchange
ASK	amplitude-shifting keying
AT	attention
ATM	asynchronous transfer mode
BCD	binary coded decimal
BIOS	basic input/output system
bps	bits per second
CASE	common applications service elements
CCITT	International Telegraph and Telephone Consultative Committee
CD	carrier detect
CPU	central processing unit
CRC	cyclic redundancy
CRT	cathode ray tube
CSDN	circuit-switched data network
CSMA	carrier sense multiple access
CSMA/CA	CSMA with collision avoidance
CSPDN	circuit-switched public data network
CTS	clear to send
DA	destination address
DAC	digital-to-analogue convertor
DARPA	Defence Advanced Research Projects Agency
dB	decibel
DC	direct current
DCD	data carrier detect
DCE	data circuit-terminating equipment
DNS	domain name server
DPSK	differential phase-shift keying
DR	dynamic range
DSP	domain specific part
DTE	data terminal equipment

DTR	data terminal ready
EaStMAN	Edinburgh/ Stirling MAN
EBCDIC	extended binary coded decimal interchange code
ENQ	enquiry
EOT	end of transmission
ETB	end of transmitted block
ETX	end of text
FAX	facsimile
FC	frame control
FCS	frame check sequence
FDDI	fibre distributed data interface
FDM	frequency division multiplexing
FEC	forward error control
FM	frequency modulation
FSK	frequency-shift keying
FTP	file transfer protocol
GFI	group format identifier
GUI	graphical user interface
HDB3	high-density bipolar code no. 3
HDLC	high-level data link control
HF	high frequency
Hz	Hertz
I/O	input/output
IA5	international alphabet no. 5
ICP	interconnection protocol
IDI	initial domain identifier
IDP	initial domain part
IEEE	Institute of Electrical and Electronic Engineers
ILD	injector laser diode
IP	internet protocol
ISDN	integrated services digital network
ISO	International Standards Organization
ITU	International Telecommunications Union
JANET	joint academic network
LAN	local area network
LAPB	link access procedure balanced
LCN	logical channel number
LED	light emitting diode
LGN	logical group number
LLC	logical link control
MAC	media access control
MAN	metropolitan area network
NAK	negative acknowledge
NSAP	network service access point

OH	off-hook
OSI	open systems interconnection
PA	point of attachment
PC	personal computer
PCM	pulse-coded modulation
PDN	public data network
PPSDN	public packet-switched data network
PSDN	packet-switched data network
PSE	packet switched exchange
PSK	phase-shift keying
PSR	packet switched router
PSTN	public-switched telephone network
QAM	quadrature amplitude modulation
RD	receive data
SAPI	service access point identifier
SD	sending data
SDLC	synchronous data link control
SEL	selector/extension local address
SNA	systems network architecture (IBM)
SNR	signal-to-noise ratio
STM	synchronous transfer mode
TCP	transmission control protocol
TDM	time-division multiplexing
TEI	terminal equipment identifier
TR	transmit data
VCI	virtual circuit identifier
WAN	wide area network
WIMPs	windows, icons, menus and pointers

Index

F

G

H

Q

R

S

Answers

1.1	B		174.113.29.57 (Type B)
1.2	D	6.1	C
1.3	C	6.2	A
1.4	D	6.3	C
1.5	B	6.4	C
1.6	A	6.5	A
1.7	D	6.6	B
1.8	A	6.7	C
1.8	C	6.8	B
1.9	A	6.15	10001100, 00001000,
1.10	B		1111101100011111
1.11	B	7.1	B
2.1	Data link, Network, Transport, Data	7.2	A
	link, Session, Physical, Prestation	7.3	D
2.2	A	7.4	A
2.3	C	7.5	C
2.4	B	7.6	B
2.5	D	7.7	C
2.6	A	7.8	B
2.7	A	7.9	A
2.8	C	8.1	D
3.1	C	8.2	C
3.2	D	8.3	C
3.3	A	8.4	A
3.4	B	8.5	C
3.5	B	8.6	C
3.6	C	8.7	B
3.7	D	8.8	C
3.8	A	8.9	A
3.9	C	8.10	C
3.10	B	8.11	samX yrreM
3.11	B	10.1	40 kHz
3.12	D	10.2	60.2 dB
3.13	C	10.3	14 bits
3.14	C	10.4	0.488 µs
3.15	B	10.5	(i) 24 channel in 125 µs gives
3.16	A		1.536 Mbps
3.17	B		(ii) 8×24+1 bits in 125 µs gives
3.18	A		1.544 Mbps
3.19	D	10.6	0.648 µs
3.28	3F5F:8850:087F	10.7	20 dB
5.1	D	11.2	0.4 ms
5.2	A	13.2	4 bits
5.3	B	13.3	00000000000 11000110010
5.4	C		01101001111 11111111111
5.5	140.113.1.9 (Type B),	13.6	Hello world
	64.125.65.233 (Type A),		*Refer to WWW page for more information.*